ONE SHOT

TREES AS OUR LAST CHANCE FOR SURVIVAL

JOHN LEARY

ONE SHOT

TREES AS OUR LAST
CHANCE FOR SURVIVAL

To the year 2117.
Hopefully by then we've straightened all this out.

CONTENTS

PART I

ONE SHOT—TIME IS RUNNING OUT

PART II

ONE WORD—TREES

PART III

ONE WORLD—LOCAL CHANGES WITH GLOBAL IMPACTS

ACKNOWLEDGMENTS

First, I would like to send my sincerest gratitude and respect to the founders of Trees for the Future, Dave and Grace Deppner, who spent years and countless hours with little pay, developing the foundation for our work today. I began my work with them and can still hear Dave's voice and lessons as I continue his work. Secondly, I would like to recognize the numerous Peace Corps Volunteers who have helped shape our organization over the decades. Many of the innovations included in our forest garden training program come from their experiences.

I would like to thank the staff at Trees for the Future who have worked tirelessly at developing our Forest Garden Program and work daily with local counterparts and farmers to implement the forest gardens in villages throughout Africa. I also want to thank our Board of Directors and support staff who secure funding, develop communications, manage business partnerships and keep excellent record of our complex accounting system. A big thanks to all our donors: individuals, businesses and foundations. I would also like to thank our editor, Doran Hunter and the staff at 1106 Design for their dedication and professional publishing services.

Above all else, many thanks to Brandy Lellou and her research assistant Rosie Phal Donaldson. Their excellent journalism and research have contributed to this work's extensive breadth and depth.

INTRODUCTION

A TALE OF FIVE FATHERS

Sixteen years ago, as a Peace Corps Volunteer, I met my friend Omar in what is known as the peanut basin of Senegal. As we sat in the barren field he inherited from his father, he showed me his ID. In French it read, *Occupation: Fisherman*. I laughed. We were one hundred miles from the ocean. Omar didn't even smile. "John," he said, "there's only death in farming."

Looking around, I couldn't argue. On his inherited land—and that which neighbored it—only a few thorny trees and dry shrubs dotted the hard, bare landscape for as far as I could see. Omar knew that peanut farming was further depleting this already exhausted land. Each year the soil required more fertilizer but yielded a smaller crop. He knew if he stayed in the village he would be farming peanuts for "peanuts"—a $200 payday that came only once a year. He also knew that along the trawled and overfished shores of Senegal, the life of a fisherman faced similar prospects. To Omar, his future in Senegal looked worse than grim, but he had what many people dream of: inherited land. He didn't need a new occupation. He needed a new way to farm.

As I traveled through the United States, talked to people and did research for this book, I discovered that Omar and other farmers in the developing world are not alone in their struggles. Both U.S. farmers and the people they are trying to feed are equally in peril. Although most of us Americans are not suffering from hunger and poverty, we are suffering from eating too much of the wrong things: grains, sugar, salt, oil and additives that have silently made their way into nearly all of our food products. Health in America has rapidly declined over the last four decades. Many of our illnesses (diabetes, high blood pressure, high cholesterol, obesity, heart disease) can be traced back to the food we eat.[1,2] We are in a farm-to-pharm cycle. For every ailment, pharmaceutical companies have a new pill. While the food we eat is harming us, the way food is grown is destroying the very land needed to grow more food.

Omar, like many farmers worldwide, was spending so much time, energy and money forcing the land to grow more in an unnatural way. He needed to listen to the land. The land wanted to return to its natural form, but the pressures were too great. He also needed to listen to himself. What did he want to do? He wanted to grow a diverse array of crops to eat and sell, but the pressures to farm like everyone else were too great. Maybe by combining what both the land *and* Omar wanted, a solution would arise.

But what would that solution look like, and how long would it take to develop it? Recently farmer, consultant and philanthropist,

Howard G. Buffett, wrote a book entitled *40 Chances*. In it he describes an "Aha!" moment he had during an agriculture training. The trainer said, "Most of you think of farming as this continual process of buying seed, planting, fertilizing, harvesting, then starting all over again."[3] In your lifetime, "if you're pretty healthy, and you're like most farmers, you're probably only going to do this about forty times. You'll get forty chances to plant your crop, adjust to what nature throws at you, and hope for the best."[4] With so many factors out of a farmer's control, including weather, insect infestation and market values, relying on one crop a year is extremely limiting—making each year a game of roulette for farmers. Furthermore, as the population grows, weather patterns become more unpredictable, the environment degrades and we discover that grain based diets are not healthy, the reliance on "forty chances" uses chance as a strategy for our continued survival. The world is out of time for chances. We are down to one shot. One shot to get it right. One shot to find a solution that can restore degraded land, stabilize soils, channel water into the ground, grow an abundance of diverse, nutrient-rich food, preserve biodiversity and end extreme poverty and hunger. But what if, within this one shot, we could create an opportunity for dozens of chances, so that crops could work together and if one failed others would be resilient? Farmers would no longer be dependent on one crop a year. This one shot that provides multiple chances is the revival of the forest garden.

Together, Omar and I began an experiment in creating a forest garden. First, we planted fast-growing trees with deep roots to improve the soil. Then we planted rows of thorny bushes around the border to keep out grazing animals and protect crops from the strong, hot winds. Once the trees were established, we intercropped a variety of vegetables and grafted fruit trees. Within four years, Omar's forest garden was generating fruits, vegetables and tree products that yielded $1,000 a year from his two acres of land, plus fruits and vegetables for his family, a sustainable supply of firewood for cooking, fodder for his goats and enough compost to naturally replenish his soils.

While Omar's forest garden was still developing, I finished my Peace Corps service, and a volunteer named Eric took my place. With Eric's assistance, Omar continued and was determined to help his neighbors transform their land by using his forest garden as a model to teach others. Eric and Omar began with Keba, a fifty-two-year-old peanut farmer with a wife and seven-year-old son. Through years of hard work and malnourishment, Keba's body had withered like the soils in his fields—both depleted from decades of peanut farming. Although Keba was determined, developing a forest garden became an exercise in resilience. During the first two years, all of Keba's trees were wiped out by locusts and goats. In Senegal, there is no crop insurance, and it was devastating to see so many months of work destroyed overnight.

While Omar and Eric continued to provide encouragement, tree seeds and technical advice to Keba in Senegal, I began working with an extraordinary man, Dave Deppner, the cofounder of Trees for the Future (TREES). Dave had been a Peace Corps Volunteer himself decades earlier. When I met him he had already planted seventeen million trees in Asia and Central America, had broken eleven bones, been bit by a snake in Indonesia, kicked by a horse in Ecuador and lost several teeth falling off the side of a mountain in Honduras. But he had done what seemed like the impossible: he had convinced small farmers throughout the world that trees were the answers to their woes.

Dave understood that agricultural land is often seen in two very different ways. On the one hand, it is seen as a green revolution, intensifying monocrop agriculture, with the idea that if the land is green, it's good. On the other hand, it's assumed that using the land for agriculture automatically means harming the environment. These two extremes lead to the unfortunate agriculture-versus-nature debate.[5] However, Dave saw a middle ground, where the quality of the environment is determined largely by the choices farmers make within their agriculture systems. These choices determine tree cover, soil stability, nutrients, seed varieties, insect control, water retention, biodiversity and the long-term productivity of the land. Dave offered these other methods to farmers as alternatives to the failing monocrop systems they had adopted from colonialists. Dave understood farmers because he was one. As a poultry farmer in Ohio for many years he witnessed the transformation of natural farms to factory farms. Later as a Peace Corps Volunteer in the Philippines working with cattle ranchers, Dave was struck by the massive deforestation and overgrazing. Dave also understood the fear of not being able to provide for a family. Although Eric, Omar and I could see that Keba's outer life was difficult, as young, single men, we had no awareness of the inner conflict he faced trying to feed his family. However, Dave, a father of five children, got it. He knew Keba's fear. Now that we are all fathers, we too understand. With this newfound awareness, I lay awake at night thinking of farmers like Keba and the desperation they must feel.

With Dave's knowledge and support, we began to transform the small Senegal project into a program. Keba's continued perseverance in developing a forest garden was rewarded. In the third year, he finally began having success.

Dave passed away in 2011, but his legacy lives on through the millions of trees he planted and through his spirit. Dave had a sincere belief about the potential of agroforestry to solve the world's problems. Now, more than a decade after Omar and I began his project, I visit him and Keba and other farmers in the region. As we walk in Keba's forest garden, massive cashew trees with thick green leaves tower above and bees hum among their flowers. I can see Keba's pride as he carefully searches the branches for the first fruit of the season. A thick living fence of over a thousand thorny trees borders his field. These trees not only protect against destructive animals but also provide jujube berries.

Keba's life has improved tenfold. He used to work in his field eight hard months each year. If the rains were good, if his horse stayed healthy and if the market cooperated, he could earn $200 a year for his labor. Now he makes that much in a single month selling the hot peppers that grow under the protection and nurture of his trees. He also fills a quarter-ton barrel with cashew nuts twice a year—giving him an additional $400. Around the cashew nut is a juicy, sweet-and-sour cashew apple. When I asked Keba how much he sells the cashew apples for, his solemn, chiseled face breaks into a smile that stretches ear to ear. "I give away more than 150 pounds of the cashew apples every season to the kids in the village." Pride, generosity, compassion and hope all beamed forth from the face of a man who once couldn't feed his own family. A man who now stands grounded and solid, flourishing and nurturing—a mirror image of his trees.

Over the years, I have spent a lot of time speaking with groups of men, women and youth with whom we work in Senegal and other countries. When I ask the men about their motivations for joining the program and planting forest gardens, they repeat the same three words: food, income and legacy. Like Omar and Keba, many of them inherited a piece of degraded farmland from their parents. They know how difficult life is trying to exist as a peanut farmer on this land. They want something of value that they can leave their children, something that can provide food and income as they enter their later years and for future generations. In a place where difficulties abound, the worst thing to lose is hope. Through this farmer-to-farmer relationship of planting forest gardens and tree crops, communities can strike a balance between forest and farmland, combining the best of both to produce useful and marketable products throughout the year. In doing so, they reignite a love of the land and cultivate cooperation, learning, teaching and hope.

You may not understand Keba's plight as a farmer, but you may share his perspective as a parent. You may not relate to surviving on a piece of degraded farmland, but you can likely empathize with the fear of not having enough to support yourself or your family. You may not wake to face hunger each day, but you may be finding it increasingly difficult to find nutritious food that keeps your family healthy and thriving.

As an individual involved in international development for more than a decade, as a friend of small farmers throughout the world and as a parent myself, I ask you to read this book with intention and a belief that you have something great to offer. You have a chance to change what and how you eat, driving farmers' choices in what and how they grow food. This will ultimately transform our food systems so that they yield not more of the same few crops, but more diverse crops that grow in unison while improving the land every season—giving farmers not forty chances in a lifetime, but forty chances every year to produce food, forage, fuel and healthy futures for their communities.

PART I

ONE SHOT
TIME IS RUNNING OUT FOR AN EFFECTIVE SOLUTION

CHAPTER 1

YESTERDAY'S PROMISE

We shall not cease from exploration
and the end of all our exploring
will be to arrive where we started
And know the place for the first time.
—T.S. ELLIOT

WALKING IN BARE FEET

Never judge a woman until you walk a day in her bare feet. Sixteen years ago, I tried. Gueye Cisse was just a baby when Senegal regained its independence from French colonization in 1960.[1] She came from a family of peanut and millet farmers who carved out the forests around Nganda, Senegal, to make a living. All she knew from her childhood was peanut farming on land that yielded less and less each year. After marriage, she moved to her husband's village of Ngodiba. The house changed, but peanut fields were the same.

In 2001, as I stepped off the plane in Dakar, our paths were about to cross. Equipped with an environmental science degree and an overdose of optimism, I was determined to help people plant trees and restore degraded lands in West Africa. My first contact in Senegal was with heat that burned my skin as if I was standing too close to a campfire. After four hours packed in the back of an overloaded taxi, I rolled into the village on a horse cart. Dust and sweat had made mud on my body. As I arrived in the village of Ngodiba, the sun was sinking into the horizon, creating silhouettes around the few baobab trees that still stood in the peanut fields that bordered the village. The smell of the wood-burning cooking fires filled the air. A rhythmic thud-thud-thud echoed throughout the village as women pounded grain with long wooden poles. As night fell, Gueye, who was to be my new host mother, returned from the field carrying a hoe, machete and a few shards of wood. She shook my soft hands, I felt her rough callouses that came from decades of hacking, chopping and digging. She spread out a mat for me to sit on the ground outside her hut, then excused herself to add wood to a cooking fire. Kids came running from all corners of the village to see me. Famished from a day of travel, I couldn't wait for dinner. Offering me everything she had, Gueye placed a large bowl on a cloth in front of me. Gueye and her five children sat down around the bowl of plain white rice with a small carrot on top. Together we shared the meager meal, each eating from the portion in front of us, with our hands. The smaller kids grabbed for the carrot immediately, while the older siblings smacked their hands. Embarrassed, Gueye pushed the carrot to my section of the bowl. Seeing the kids longing for it, I flicked it back to them, and so it continued. It was there, during my first night in the village, that I began to learn about hunger, up close, by playing finger football with a carrot in a rice bowl. After dinner we began to hear drumming and singing. "Our neighbor is having a wedding," Gueye explained. As the drummers played a rhythm that matched the mood, a sea of colorfully dressed men, women and children crowded the groom's dirt courtyard, spilling out into the roads and surrounding compounds. Soon everyone danced to the rhythm. The rising moon cast a glow on the village, lighting up the bright smiles of the people in my new home. Despite the heat, sweat and mosquitoes that had begun their attack, I felt like all was right in the world. I breathed it all in as my stomach rumbled.

The first week I struggled to keep up. At sunrise, I attempted to carry water from the well to Gueye's house. I could fill the bucket but couldn't lift it to my head without spilling the water. Once upright, I looked so shaky that the women were relieved to see me put the bucket back down for Gueye to carry home.

After our trip to the well, we walked a mile to Gueye's peanut field, picking up small shards of wood or brush to use for cooking later in the evening. Under an unforgiving sun, we toiled all day weeding and clearing the land of unwanted growth, chopping and hacking anything that wasn't a peanut plant: new tree seedlings, grasses and weeds. At the end of the day, the "debris" was gathered in piles to dry and eventually be burned. While walking the long way home we looked for more pieces of wood and brush to use as firewood. Our dinner of boiled millet with leaf sauce was for bare sustenance only. It was followed by the only pleasure of the day: sweet tea. As we fell asleep under the stars to the smell of burning charcoal, I couldn't help but think of the irony—I had come to plant trees, but my only solace was the tea heated by burning them. Each day I burned more calories than I consumed, and soon I was too focused on surviving to think about planting trees. My vision of a grassroots movement had been very different from reality. In the village of Ngodiba there was no grass. If we were lucky we ate roots, and the heat left very little energy for movement. Everything we did, it seemed, was to remove trees from the land, yet I became too tired to plant them.

Gueye and her neighbors are not alone. For more than a billion people, the only life they've known is quickly coming to an end as soils deplete, water becomes scarce and the last trees are cut down.[2] The connection between trees and agriculture is not intuitive until you look closely. Over the last forty years, entire countries have been stripped of a primary natural resource: trees. Three quarters of this deforestation occurred to make room for agriculture.[3] Although there is a constant pressure to trade more forests for food, this exchange has not resulted in sustainably producing more food. Instead it has left a wake of hunger and devastation across the developing tropics, creating a ripple effect that impacts soils, biodiversity, water, the climate and people. In Part II of this book I unpack how our current agriculture system is contributing to seven of the world's most pressing challenges. This is not only a problem for the developing world; farmers in the United States and other developed nations are also experiencing the problem of how to feed ourselves without killing ourselves.

THE GREAT CORN CIRCUS

Six thousand miles west of Gueye's village, on three circles underneath the big sky of Colorado, Betty, who wishes to remain anonymous, is not hungry. Thanks to the American farm credit and insurance system, she is not living in poverty. The irrigation water on their family farm is pumped with electricity from deep wells, and she cooks with piped-in propane gas, not wood. Despite the conveniences of Western civilization, though, Betty has discovered first-hand the results of failed farming systems. For forty-five years, on four hundred acres of irrigated land, Betty and her husband have attempted a balancing act between heat and sunlight, moisture and rain.[4] Across the U.S. Corn Belt 340,000 other farmers are orchestrating the same amazing feat—the production of a third of the world's corn in five short months.[5,6]

Some years, the ring master is good to them. Early rains and mild temperatures provide perfect growing conditions, allowing the pollen to find its way to the silks and the corn to begin to fill in with big fat kernels. However, many years they are not so lucky; the devastation of ruined corn crops brings even the most seasoned farmers to their knees. The year 2012 was one of those years. "We've been farming corn for forty-three years and that was the first time that we had some fields that didn't pollinate. They only had a few kernels on the cob, and those were irrigated fields," Betty stated during an interview in 2014. "From the outside the plants looked great, but the corn just didn't develop."[7]

Corn has an interesting reproductive system. Those pesky silks that we find so difficult to remove are actually the corn's female reproductive organ (there is one silk for each kernel). The corn's male reproductive organ is the tassel on the top of the stalk. When the wind blows, the tassel's pollen is blown and caught by the silks (a separate mating occurs to grow each kernel of corn—with about eight hundred kernels on each corn cob). However, corn is sensitive to extended periods of heat above 95 degrees.[8,9]

"It was so hot for such a long time, even the nights didn't cool down. It didn't matter how much water we pumped from the wells," insisted Betty.[10] The most devastating drought in fifty years, combined with extreme heat, prevented pollination. The result was rows and rows of stalks with empty cobs across sixteen states, leaving a wake of deprivation not only for corn growers, but all industries that depended on them.[11]

One year's lost crop is not the only problem Betty and her husband face. At seventy-two years of age they cannot retire. They have large debts, their wells are in jeopardy of being turned off (due to centuries-old water-rights disputes between states), and each year fertilizer, seeds and herbicides are getting more and more expensive. In 2013, the weather cooperated and they had a bumper crop—the best they'd had in years—and the corn market was up. But after the cost of seed, energy to pump water, fertilizer, herbicide, insecticide, insurance and interest on their loans, they only made $34,000 for the entire year's work. Betty and her husband don't take vacations, they work an average of ten hour days, six days a week. If you paid them hourly they'd make $5.60/hour, well below the minimum wage. "We use to say it was worth it for the lifestyle, but I'm not sure anymore," Betty said.[12]

Though continents apart, Gueye in Senegal and Betty in Colorado face similar problems: poor diet, lack of opportunities, little return on their investment, depleting soils, and constant risk. Throughout the world, the problems that nearly every farmer faces can be tied to one thing: the failure of one-crop systems.

PLAN A—CASH CROPS: NO CASH, NO CROPS

The U.S. began Plan A with the move west, clearing grasslands and forests to plant crops (wheat, corn, cotton, millet) but by the 1930s the results were catastrophic. Severe drought, strong winds and millions of acres of plowed-up grassland led to massive dust storms and the loss of twelve inches of fertile topsoil. In 1950, the government stepped in, creating the Conservation Reserve Program (CRP), which payed farmers to plant their farmland back to grassland and preserve it. However, by the early 70s the world grain market was exploding and the strategy changed again. The Department of Agriculture developed a new motto: "Get big or get out." Getting big required investments in more land, fertilizers, insecticides and large machinery. Farmers took out massive loans and crop insurance companies rose up to insure the crops that failed. Yields and profits grew for almost a decade. Then in 1979, the Soviet Union invaded Afghanistan. As punishment, the United States imposed a grain embargo on the Soviet Union. This led to vast U.S. surpluses and plummeting market values in the early 80s, causing many small farmers to go bankrupt. In less than fifty years, between 1940 and 1985, over four million small farms failed, opening the door for large agribusiness to impose the industrialized farming we have today.[13]

During this rise and fall of the small farmer in the U.S., small farmers in Africa were getting bullied by other forces. Under colonial rule the plantation system of farming was widely introduced to grow large quantities of cash crops, employing cheap (often forced) African labor for export to European countries. Many African farmers were pushed off their fertile lands to marginal lands and encouraged to grow cash crops to generate exports and pay for high taxes.[14] In many countries colonial law stated that trees were the property of the government. Although this law was intended to protect forests, it had the opposite effect as farmers removed trees from their farmland to prevent their land from being policed.[15] These laws began the separation of people from the trees. In the 50s and 60s, as African countries gained independence, people returned to their lands but the plantation systems had caused severe environmental degradation and depleted soils.[16] Thus began the phase of neocolonialism throughout the developing world, as developed countries stepped in with aid packages and loans to showcase how agriculture was being done in the West. This led to the first Green Revolution, in which Nobel Peace Prize winner Norman Borlaug and the Rockefeller Foundation introduced high yield wheat and rice varieties to farmers in Mexico, Central America, India and the Philippines. Forest, pasture and multi-cropped land was cleared to plant these monocrops that "provided bread for the world" but ultimately depleted soils and shackled farmers to a lifetime of paying for seeds and fertilizer. In trying to solve the hunger crisis of his time, Dr. Borlaug unknowingly created a grain dominance within the world's food supply, "displaced smaller farmers, encouraged overreliance on chemicals and paved the way for greater corporate control of agriculture."[17] This first Green Revolution skipped over Africa. In the 70s and 80s when drought and massive famines plagued the continent, wheat and rice were brought in as food aid. Thus, beginning the Western perception of "starving Africans" that we still associate with the continent today.[18] With all these pressures, many rural farmers across the developing tropics from Manila to Managua turned to Plan B.

PLAN B—MASS RURAL EXODUS

About the time that cellular communications and the Internet were taking off, another phenomenon was occurring that would vastly impact developing nations: a global mass exodus from rural areas, as farmers and villagers moved from their degraded lands to cities in search of work, food and water. This trend conjures up images similar to those from the United States decades earlier, so well-described in Steinbeck's *Grapes of Wrath*.[19] Most migrants arrived only to find conditions far worse than those they left behind. The cities they migrated to had little or no infrastructure to support a population influx, and thus new arrivals created shanty towns in flood zones, along coastal areas and on unstable hillsides. This rural exodus to nearby urban areas has quadrupled many urban populations in the last three to four decades.[20] Coupled with a lack of infrastructure and no jobs, Plan B has become a nightmare. Plan B has left people living under tarps, near garbage dumps and without access to toilets or water. It has resulted in the increased spread of diseases and violence. It has driven children to beg in front of boutiques packed with processed cookies, chips and canned meats that they can never afford. It has created an hour-by-hour struggle that puts people at major risk. During the monsoon and hurricane seasons, people living in shanty towns, in flood plains and on unstable hillsides have nowhere to go as their makeshift homes become flooded. Many countries, particularly the Philippines, India, Indonesia and Nigeria have become hotspots for human trafficking, because people, whose problems first began with the disappearance of trees, have been left vulnerable and desperate.[21] Daily, thousands of people risk their lives on dangerous journeys across deserts and oceans, on foot and in small boats, in search of hope and opportunity in other countries.[22] Many do not make it. Those who do are met with resentment from country nationals and often live in very difficult situations without proper documentation. In developing nations, Plan A resulted in loss of land, but Plan B resulted in the worst thing a person can lose—hope.

A SLOW DEATH

Plan B played out differently in the U.S. At the end of 1985, as farmers lost their farms at a rate of 250 farms per hour, there were no famines.[23] American farmers had become so good at growing grains, there was an enormous surplus. To increase the market for these grains, chemical companies transformed surplus grains into the sugars and additives that became new ingredients for the processed foods and beverages that have become the staples of our diet today.[24] Unlike the mass exodus in developing countries, in the U.S., rural to urban migration occurred slowly as the next generation moved from small towns to urban areas. However, the psychological legacy of loss remains today.[25] Farmers love what they do, they love the land. They have a physical need to produce and work with their hands. The loss of that love, unemployment and underemployment have resulted in the pervasive social ills that continue to plague rural areas: increased alcoholism, drug use, domestic abuse and suicide.[26,27] Instead of working the land, the descendants of failed farms who stayed in small towns often work on large feedlots or in fast food, serving their neighbors chemically treated beef that was fattened on genetically modified corn. Instead of eating fresh food produced from local farms, they are drinking beverages loaded with high fructose corn syrup and eating frozen, ready-made meals, chips, cookies, crackers and other processed foods trucked in from food factories. They are not alone. On average, we Americans get only 1 percent of our daily calories from vegetables, while 60 percent of our calories come from "ultra-processed" foods consisting of ingredients we could not find in a kitchen or grow on a farm. This has led to widespread obesity and a plethora of health problems.[28] Although doctors stress that the key to health is avoiding excessive sugar, salt, carbohydrates and gluten, our food systems continue to produce these ingredients in abundance and hide them in the majority of food products available.

Whether in Senegal or Colorado, the loss of small farmers has resulted in the loss of our connection to the land, loss of hope, loss of health, and loss of dignity. It is in the interest of our future that we reconnect with our food, begin to understand the land that it comes from, what goes into it, what it takes to grow and the impact it has on our lives and on the lives of others. By better understanding the sources of our food today, we can make choices about the food of tomorrow. We can avoid being the victims of yesterday's promise and create a new plan, a Plan C, that equally takes into account all of our needs and the needs of the soil beneath our feet. In understanding the connection between food and the natural state of the land, we can develop a Plan C that sustainably ends hunger and poverty while improving the health and prosperity of people and the land. To do this will require rethinking agriculture. Not just what we grow, but how we grow it, and in this, trees make all the difference.

CHAPTER 2

TOMORROW'S
UNCERTAINTY

*In every scheme that worsens the position of the poor,
it is the poor who are invoked as beneficiaries.*

— DR. VANDANA SHIVA
physicist, author, environmental
activist

WEB OF INTERDEPENDENT NEEDS

If farmers in Africa and the United States are experiencing much of the same failed agricultural problems, why do we feel so far removed from them? Perhaps because we are. I live in a subdivision off of *Quince Orchard Road*. Quince is a fruit that can be found at the supermarket about two weeks out of the year. However, like many subdivisions that bear names such as *Farm's View* or *Green Acres*, there hasn't been a quince orchard within 250 miles of my home for decades. For many subdivisions that sprawl out from city centers, these quaint names are all that remains of the area's farming history. Like myself, many American urban dwellers are two to five generations removed from the land. Meaning Betty's four hundred acres in eastern Colorado may as well be Gueye's five acres in Senegal; we are equally as detached from one as we are from the other. Although agri-culture is no longer a part of our culture, our need for nourishment ultimately binds us all to the land. Perhaps in the U.S. this is most true with corn. We only eat sweet corn a few times a year, unaware that corn has many forms. Corn is crucial not only to backyard barbecues, but also to flu vaccines, DNA coding, cosmetics and 75 percent of all our food products.[1] As a grain, a starch, a sugar, an oil, an additive, a primary source of animal feed and a base for 10 percent of all fuel, corn growing is ten times more lucrative than gold mining in the U.S.[2] The absence of agri-culture from our culture has created a blind spot in our understanding of where and how our food is grown and processed. This blind spot allowed for the slow and silent introduction of non-food additives and genetically modified, processed corn derivatives into nearly all our food products. Therefore, whether we eat it, put it on our faces, feed it to our dogs or put it in our fuel tanks, unraveling corn from our lives is becoming increasingly impossible.

Likewise, the nutritional, environmental and economic issues facing the world are as intertwined as corn is in our products. Globalization has brought forth a Celtic knot of interlocked issues making it difficult to identify the most pressing local, regional and global problems and choose what to address first. In the fall of 2015, the United Nations Development Program

(UNDP) announced their 17 Sustainable Development Goals (SDGs).[3] In a grand campaign to create global awareness of these goals, they advertised through the web, radio, television and movie theaters using top-name musicians, actors and sports stars. This campaign aimed to reach seven billion people. Nearly every living person across the globe had the problems laid out in front of them: from poverty and hunger to deforestation and water pollution, from refugees and human trafficking to lack of education and climate change.[4] However, recognizing the problems is only a tiny step. When we begin to investigate solutions, more questions arise. What is the most pressing issue to tackle first? How can we solve one problem without addressing the others? What is the root problem we must solve?

Between 2000 and 2015, billions of dollars were poured into addressing the Millennium Development Goals (MGDs)[5], the precursor to the SDGs, but the problems have only persisted. Within the development community frustration has grown as well. Each agency or organization has an area of focus: nutrition, education, economic reform, water access, health and sanitation, environmental conservation and many others. Though each focus area is represented in the SGDs, when international development professionals begin working on an issue, they discover that Celtic knot. Think of it as a large ball of spaghetti, problems so tightly interwoven that one cannot be touched without encountering another. Nutrition cannot be solved without access to more diverse foods. More diverse foods cannot be grown without better soils and protection of farmland from grazing animals. Farmland cannot be protected without wood for fences. Wood cannot be found without cutting trees and the last remaining trees are being cut to cook the meager meals that have little or no nutrition. These interwoven problems also make concerned citizens and donors apprehensive and wary. What programs should they invest in? Where can they have a sustainable impact? Again, we return to the question: what is the root problem we must solve? There are many initiatives that address symptoms, yet hundreds of millions of subsistence farming families continue to live and suffer in extreme poverty and hunger. The root problem we must solve is not climate change nor the loss of biodiversity or hunger. These are all symptoms of the collapse we are headed toward unless we change the way we grow our food.

The root problem is that the way we are growing food is destroying the very land we need to continue growing food. In an effort to feed the world, forest lands are cut and burned, grasslands overgrazed and plowed up and mangroves destroyed. This eliminates trees, degrades soils, pollutes water and kills biodiversity—leaving us land that can no longer support human life—while increasing greenhouse gases. To ensure high yields, monocrops require expensive, environmentally polluting inputs to maintain productivity, including genetically modified seeds, chemical fertilizers, herbicides and pesticides. In a farming system that deploys these new inputs, the input providers make all the profits while farmers bear all the risk. While watching their farms slowly die, farmers using these monocropping systems also suffer from nutrient-deficient diets that lead to a host of health problems. Hunger, poverty and desperation lead farmers to further deforest and degrade lands in their struggle to survive.

The chapters in Part II of this book unpack seven of the world's most pressing challenges and demonstrate how agriculture is at the center of the Celtic knot. Trees are the missing link to solving these interwoven problems. Each chapter reveals how trees, planted in forest gardens, are the key to optimizing and maximizing farming output, restoring soil nutrients, managing water, ensuring biodiversity, improving health, and sustainably lifting a billion people out of extreme poverty.

Due to destructive agriculture systems, the world is facing an environmental and humanitarian crisis. One that began with the removal of trees from the land and continues. Humanity cannot simply adjust to this dire situation by adding more chemicals to the land, or moving to urban areas. We must find a solution that both meets the world's growing nutritional needs and places agriculture on an environmentally sustainable foundation. A crucial component of this solution is converting industrial agriculture land to forest gardens.

DOWN THE RABBIT HOLE

Trees are much more than just organisms that suck in CO_2 and create oxygen; they are truly magnificent creatures with the power to reverse much of the environmental destruction humans have caused. A mother tree is not just the largest organism on a landscape but a steward of the land. Not only does she rain down seeds all her life, she recognizes her own seedlings and those of other trees, both of her own and of different species.[6] She'll even retract her roots so as to not compete with her babies. What is really mindboggling is that mother trees use the healthy network of mycorrhizae fungi that form a web across the understory of a forest—like a scene out of the movie *Avatar,* these trees send messages back and forth and through this web. In healthy soil the mother tree passes nutrients, moisture and even alleles (antibiotics) to her seedlings. And she will send messages like "forest fire" so that the babies can prepare.[7] Mother nature heals herself. She is coded with the plan to go from grasses to bushes to rich diverse forests. Things want to grow up, if left alone.

But we don't leave anything alone. Humans have transformed the global environment in the last ten thousand years, with much of the degradation happening in the last hundred years. To feed our growing population, we have spent the last hundred years using fossil fuels in the forms of herbicides, tractors, machinery, pesticides and even just burning to keep mother nature from doing what she wants to do.

While American farmers tend to measure their topsoil in inches, farmers I work with in Africa measure it in millimeters, if anything is left. The dire state of soils across Africa resulted not only from agricultural intensification, but also from overgrazing by roaming livestock who eat all vegetation (including every tree seedling) and compact the soil. At some point the land loses its ability to heal itself. Once the forests and trees are wiped out, the soils erode, the watersheds dry up and nothing will grow, communities face collapse. To survive they either migrate to urban slums or cross deserts and oceans looking for a better life. But even in urban areas people must eat and we are rapidly losing what we need to grow food: nutrients in soil, clean water and even an element of chemical fertilizer (phosphate). At the current rate, by the end of the century, every landscape on earth will have lost its trees, bushes, bacteria and other microbes needed to naturally restore soils for productive agriculture. We will have lost our one shot to get it right and we will see our global crisis amplify.

Although I consider myself an optimist, there is very little glow on the horizon. At the dawn of agriculture there were five million people on the planet. By 1970 there were 3.5 billion and since then the population has more than doubled to 7.5 billion.[8] By the time my son is my age there will be almost ten billion people on the planet. What does that mean exactly? Simply put, it means more competition over fewer resources.

Today, eight-hundred million people suffer from not having enough food, two billion more suffer from not having enough of the right food, and another two billion suffer from having too much of the wrong food.[9] Thus, the majority of the world's population, including Americans, are not eating foods that maintain their health. Due to lack of nutrients, fifty percent of children in low-income countries show growth, cognitive and developmental stunting by the age of two. Children rarely catch up after that age.[10] Girls and women walk up to eight hours a day in Ethiopia to find water and firewood.[11] Grazing herds of animals turn millions of acres of grasslands to deserts per year.[12] In India forty-one farmers commit suicide each day due to hopelessness and debt.[13] Soil erosion has made a third of arable land unproductive in just four decades.[14] In Haiti less than half of all children are able to attend school.[15] Largely due to poverty, human trafficking is the third largest and fastest growing criminal industry in the world.[16] Thawing permafrost in Siberia is not only releasing greenhouse gases, but also reviving anthrax bacteria from a previously frozen reindeer carcass.[17] It is quite likely that before I retire the earth will be 2 degrees warmer, and sea levels will rise at least one foot, transforming our coastal cities in ways we have yet to imagine.[18]

With these looming disasters, we are out of time for experiments. The Celtic knot cannot be unraveled; it must be tackled in place with a new plan, a Plan C. The United Nations Sustainable Development Goal for hunger states, "A profound change of the global food and agriculture system is needed if we are to nourish today's 795 million hungry and the additional 2 billion people expected by 2050."[19] What kind of profound change will we adopt? Will we adopt a sustainable change that provides a variety of nutrient-rich food, improves the lives of the world's poor and ensures that farmers and the earth can produce food in the future? Or will we adopt an increasingly industrial change that fills our stomachs with nutrient-deficient grain products, lines the pockets of large agribusiness and destroys the diversity we will need in the future? We struggle to answer this as a society because corporate interests are so interwoven in our food systems. However, it really is a no-brainer. Our question then is not *what* kind of change we will adopt but *how* we will make this change.

If today's mega farms and even small U.S. farms, like that Betty operates, would run at a loss were it not for insurance and government subsidies, what kind of profound change can small farmers like Gueye Cisse implement on two to five acres in Senegal? For decades, agriculture technologies including chemical fertilizers, insecticides, advanced machinery and genetically modified (GM) seeds have promised to increase productivity and end food scarcity. Instead they have led to land degradation, debt and dependency. Like any big challenge, it is human nature to think that this problem requires a big solution. And it does. But instead of a handful of agribusinesses selling inputs to billions of small and large farmers to raise the yields of a few grain and cash crops, Plan C requires billions of small farmers, like Gueye, taking control of their one to ten acres to sustainably grow the wide variety of diverse crops that can nourish their communities, restore their lands and lift them out of extreme poverty.

PLAN C—A DISRUPTIVE INNOVATION

This profound change does not happen overnight. It starts with what is known in the tech and business world as a *disruptive innovation*. A disruptive innovation is a solution to an overlooked problem that begins to get the ball rolling in a new direction.[20] In the case of small rural farmers, they have many overlooked problems (poor soils, no protection of crops from herds of grazing animals, little forage for their own grazing animals and lack of firewood for cooking). When we evaluate these overlooked problems, we see the potential for a disruptive innovation. Our disruptive innovation is the forest garden. Forest gardening is not a new idea; it has been practiced for centuries by indigenous farmers, and many forest gardens exist throughout the world. The largest grouping of forest gardens is in one of India's most densely populated states, Kerala.[21] Kerala has 3.5 million forest gardens, each about one acre in size, that provide a surplus of wood, fruits, vegetables, steady income, health and environmental prosperity for residents.[22, 23] However, the methodology for developing forest gardens on previously mono-cropped, degraded, barren land is new. The forest garden methodology of *protect*, *diversify* and *optimize* is an alternative to the current farming systems that *clear*, *intensify* and *degrade*. The difference is fundamental. Emphasis on intensifying—increasing yields—only results in making a destructive agriculture system more efficient at producing one thing. The forest garden involves designing and implementing an agriculture system where all the inputs, products, by-products and processes improve the health of the land, those who farm it and the communities that depend on it. The forest garden provides for people and the planet, allowing a profitable system to be developed that addresses the web of interdependent needs. It improves family nutrition through the harvest of a multitude of crops throughout the year. It provides protection of crops through the planting of a three-layer living fence that borders the field. The living fence provides a sustainable supply of firewood for cooking, berries and nuts to eat, leaves that make an excellent source of animal fodder and compost that improves soil nutrients. The roots of the trees in the forest garden funnel water back into the ground, strengthening the field's resistance to both flooding and drought. Byproducts from the forest garden can be sold for income to buy clothes and materials, and send children to school. By working with each farmer to create a vision, their unique forest garden can provide a diverse array of opportunities for fulfillment and the chance to leave behind a legacy, something of value for their children to inherit. Not a dry, barren, degraded piece of land, but a lush forest garden that produces a variety of vegetables, fruits and trees, wood, forage, water and nutrient-rich soil. The forest garden methodology alleviates the fear of tomorrow's uncertainty and replaces it with hope. It removes the weight of dependence, replacing it with the dignity of self-reliance. The forest garden creates a newfound belief that the hard work of farming is about more than high yields—it's the key to solving all the problems of the interdependent web.

PART II

ONE WORD

TREES

CHAPTER 3

HOOVES AND AXES BREAK THE LAND

DEFORESTATION/DESERTIFICATION

*We want it now—and if it makes money now
it's a good idea. But if the things we're doing are
going to mess up the future it wasn't a good idea.
Don't deal on the moment. Take the long-term
look at things. It's important that we do the right
thing by the soil and the climate. History is of
value only if you learn from it.*

—WAYNE LEWIS
Dust Bowl survivor

WHERE THE GOATS AND CATTLE ROAM—FROM NEBRASKA TO BABYLON TO TIMBUKTU

Following the Civil War there was an abundance of abandoned cattle in the southern United States. In search of new markets for these cattle in the North, Captain John Lytle and several cowboys drove three thousand five hundred cattle and a few hundred horses from southern Texas to the railway head in Fort Robinson, Nebraska. Soon others followed, establishing the Great Western Cattle Trail. For nearly twenty years the trail was used to move six million cattle and one million horses north, leaving a trail of destruction in their path. By the early 1900s there were nearly twenty million cattle and over twenty-five million sheep grazing over seventeen western states.[1] These great herds of roaming animals degraded the western grasslands of the United States, creating what Victorian economist William Forster Lloyd referred to as the "tragedy of the commons."[2] Alongside this tragedy of the commons, another tragedy was occurring, the "tragedy of conversion," otherwise known as "The Great Plow-Up." Between 1910 and 1930, 5.2 million acres of pristine grasslands were converted to wheat fields. Coupled with an eight-year drought, the degradation of grasslands by roaming herds of animals and conversion of grasslands to farmland led to desert-like conditions, resulting in the Dust Bowl of the 1930s.[3] Today, this should serve as a wakeup call for how quickly grazing and plowing can destroy a virgin ecosystem. In just three decades, a population of less than thirty million people[4] and forty-five million animals transformed the pristine untouched grasslands of the West into a desert.

The tragedy of the commons and the tragedy of conversion began long before there were names for them, and they continue today throughout Africa and much of the developing world. We need only to look to the Near and Middle Eastern Countries of Afghanistan, Pakistan, Turkey, Syria, Jordan, Yemen, Saudi

Arabia, Iraq and Iran for a visual representation of what Africa will become if this trend is left unchecked. Within the time-frame of documented history, large forests existed in Near and Middle Eastern countries. Ancient cedar, pine, fir, juniper and frankincense forests were decimated by unselective chopping, burning and overuse. In a display of wealth and power, the cedar forests were used to construct massive temples and palaces in the empires of the Fertile Crescent. [5] Likewise, it is no coincidence that these countries that were the birthplace of livestock inten-sification now suffer from desert-like conditions. Unrestricted grazing of cattle, sheep and goats prevented the land's natural regeneration of vegetation. [6]

Without vegetative cover, soils become fragile and dust like. Roots in the soil are similar to rebar in concrete: they bind it together. If a concrete structure is built without rebar, it will soon begin to crumble and become a pile of rubble. Similarly, soil without roots begins to erode. When the wind blows and the water flows, the soil goes with it. As with the Dust Bowl centuries later, man started the disaster long ago in the Near and Middle East, and nature took over.

From the Middle East, livestock domestication extended across Africa, degrading the continent in ways that herds of elephants, gazelles and wildebeests never did. [7] Today, three-quarters of all livestock in Africa is found in drylands areas. [8] However, drylands are not the only areas that are becoming desert-like. Although we often associate desertification with blowing sand dunes and massive heat, desertification does not have an altitude or climate. Just as loss of grasslands and trees led to desert-like conditions on the western plains of the United States and across the Middle East, loss of trees and grasslands from Senegal to Ethiopia is creating deserts across the Sahel in what was once forests and savanna lands.

Of all the degraded land in the world:[9]

35% is due to overgrazing

30% is due to deforestation

28% is due to agriculture

In 1977, out of concern for land degradation in the Sahel, the United Nations held the first Conference on Desertification (UNCOD), where it was declared that the Sahara Desert was advancing southward at a rate of about four miles per year.[10] In light of this shocking news, a plan was developed to plant a 4,350 mile-long "Great Green Wall" across Africa to prevent the Sahara from moving further south.[11] If the desert was rapidly advancing south, this grand scheme may have been a solution. However, the desert was not and is not actually moving south. Rather land degradation is radiating out from villages across the Sahel for the same reasons it did in the U.S. nearly a century ago. Large herds of roaming goats, sheep and cattle are eating everything in sight, leaving dry, barren land, marked only by a few trees in their wake. As the population (of people and animals) grows, the circles of degraded land around villages grow. The remaining trees are being cut for firewood and to make fencing to keep grazing animals from eating crops. The loss of these trees' roots further destabilizes the eroding soils. Eroded soils are unable to naturally regenerate grass, and the cycle of land degradation continues to expand. During the thirty-year period in which money was being poured into planting the Great Green Wall between the Sahel and the Sahara, 80 percent of the trees died, while the human population doubled and the number of grazing animals nearly tripled.[11,12,13] If the wall had succeeded there would now be a green band of trees between the existing Sahara Desert and the rapidly developing desert of the Sahel—a reminder that planting trees alone is not a solution. Trees must be integrated within rural communities, not apart from them; we cannot separate the people from the trees.

In the attempt to create a Great Green Wall, most of the trees perished because there was no one to look after them. In harsh, desert-like conditions, young trees need to be nurtured. However, people who are already living on the edge of poverty will only exert the effort to nurture trees if they can see a direct, tangible benefit. They will not plant trees for the sake of trees; they will only plant trees to save themselves.

HOW MANY LEGS DOES YOUR INVESTMENT HAVE?

To understand why animals are ubiquitous and abundant throughout Africa and many developing nations, we need to understand their role in society, not only as a source of protein, but as a secure investment. In urban areas of the United States and other developed countries, many of us never think of investing in something with legs. However, in rural areas worldwide, investments with legs are a portion of nearly every farmer's income. In developed countries, these investments are sold on the commodities market as futures and thus their value is somewhat protected from extreme price fluctuations. However, if we lived in a rural area of the developing world, our investment would eat a dwindling grass supply and its life and ours would be co-dependent. If our lives depended on the life of this investment, how far would we go to protect it?

For the rural poor in Africa there are no futures markets, no credit cards, no access to lending institutions and no insurance. Their investments are the cattle, goats, sheep or camels they own. Because of unfair banking and lending policies, even most wealthy residents distrust banks and prefer to put their money in four-legged accounts. Looking at it from their point of view, it may be hard to argue. Animals provide a return on investment in the form of milk and eggs. They can act as a form of insurance when disaster strikes or crops fail. The investment grows aggressively with each offspring, and animals ensure that money isn't unnecessarily spent. If you have a cow as a source of future income, you can't sell a portion of it; you either have it or you don't.

However, animals are very high maintenance—they need food, water and protection. If you owned one four-legged bank account (a cow for instance) it would require more than thirty-five pounds of food per day to remain stable. These animals are not fenced in or fed corn, soybeans and alfalfa; they roam degraded lands eating dried grass with the nutritional value of cardboard. During the dry season, free-ranging livestock are desperate for food. They often burn more calories searching for food than they restore from eating it. They can lose up to a quarter of their weight and often quit producing milk. Although cattle are extremely destructive, they primarily graze on grass and leave some vegetation behind. Goats, however, with their protruding teeth can eat every inch of grass, seedlings and anything in sight. They can even be seen climbing trees to eat the leaves. Alongside the problem of overgrazing is the growing demand for meat and animal products. This demand will continue to grow as the population grows, urbanizes and adopts a Western diet. Within the next decade, global livestock production for meat and milk is expected to grow by more than 70 percent to meet rising demands.[14] However, due to land degradation, there is little room for pasture expansion. In more developed countries this deficit in available pastureland will be supplemented with grains.[15] Already 36 percent of all agriculture land is used to grow grain for animal feed, primarily corn and soybeans.[16] This will divert even more agricultural production away from crops for human consumption and towards livestock feed.[17] Thus, about the time that the population reaches eight billion people, meeting the increase in demand for food will require growing more, on land that is quickly losing its ability to grow anything.

This is an ethical and environmental challenge, but like all problems in the developing world it is not a singular issue. To solve it requires finding alternatives: an alternative to destructive overgrazing, alternative fencing, alternative feeds, alternative income, alternative insurance and alternative investments. There is a solution for these environmental and economic woes—a solution that grows from one thing—planting trees in the form of forest gardens. Trees, with their renewable supply of fruit, nuts, leaves, wood and other valuable byproducts essentially become fast-growing bank accounts for farmers in developing countries, bank accounts with roots instead of legs.

GROWING FENCES AND FORAGE

Sometimes alternatives can come from the top down, as they did in response to the 1930s Dust Bowl. With the passing of the Taylor Grazing Act, individual ranchers could receive ten-year private grazing leases on public land. This gave them an incentive to manage their pastures, knowing that any grass left standing wouldn't simply be eaten by a neighbor's livestock.[18]

However, in Africa, despite the desperate land degradation situation, federal governments are not stepping in to divide up common land for private lease. Most privatization of public land involves large corporation land grabs.[19] For small farmers, in rural areas of the developing world, positive change is not going to come from the top; it must involve community organizing and their own initiative from the bottom.

From experience, we know that all initiatives that start from the bottom must start small. When a community starts small, they do not aspire to plant a four-thousand-mile Great Green Wall. They follow the old Wolof saying: "slow and steady catches the monkey in the forest."[20] Over the years we've learned that by helping farmers plant three staggered rows of trees that form Green Walls around their small plots of degraded farmland, they can slowly and steadily reverse much of the past damage from overgrazing while producing healthier animals and making more income. These rows of trees serve as dual protection—acting as a living fence to protect crops from grazing animals, and a wind break to protect the field from the harsh elements. This Green

Wall disrupts the cycle of deforestation and overgrazing while increasing the productivity of the field. By planting rows of trees on the most underutilized real estate of a field (the perimeter), farmers begin the process (protect, diversify, optimize) of converting monocultures and cash crop fields into forest gardens.[21]

The outer-most row of a farmer's Green Wall of trees is comprised of thorny trees or bushes planted very close together to form a thick hedge that even goats cannot penetrate. In Africa, we frequently use *Zizyphus mauritiana* (the Jujube) and *Dovyalis* (the Kai apple) because both thorny species have a valuable vitamin-C rich fruit that can be sold in markets or consumed by the family.[22]

The middle row tends to consist of fast-growing but sturdy trees that give the living fence the structural integrity required to support the many thorny branches that are eventually woven among each other. We frequently use trees such as *Moringa oleifera*, *Jatropha curcas* and *Acacia nilotica* to provide this line of structural support. The leaves of moringa trees are high in protein and vitamins and can be added to any meal to increase nutrients. Jatropha trees produce oil rich seeds that can be used for biofuels. The acacia is a highly prized local tree that is nitrogen fixing, medicinal and used for tanning leather.[23]

Farmers often make different decisions as to which trees to plant on the inner lining of the Green Wall that surrounds their field. We recommend planting tall, fast-growing trees such as *Leucaena*, *Cassia* and *Glyricidia*, to form a multi-purpose windbreak. The branches of these fast-growing trees can be selectively harvested to provide a sustainable supply of firewood for cooking. Their leaves also make excellent forage for livestock across

Africa. Most farmers are surprised to learn that some of the best forage grows on trees. Cows, goats and sheep enjoy the protein and nutrient-rich leaves of numerous trees. By growing trees for forage, farmers can keep their animals from rummaging the countryside while increasing their weight and milk production. Farmers in areas better protected from fierce winds, or those who do not have animals, often opt to plant rows of nitrogen-fixing *Cajanus cajan* (pigeon peas) for the inner row. This allows them to harvest dozens of pounds of protein-rich pigeon peas from their living fences twice a year. By planting a Green Wall, farmers can produce more on a field's previously neglected real estate (its border) than they could before on the entire field.[24]

Our Green Wall (living fence) provides forage, wood, fencing and food for the family while continuously building soil stability and preventing erosion. While the living fence is growing, farmers can continue to grow their traditional monocultures or cash crops in the protection of these trees. Once the living fence is established, farmers can begin to convert their fields into forest gardens by growing a variety of trees and plants that provide sustainable products they can eat, use and sell throughout the year. By protecting fields with living fences and growing diverse resources on one plot of land, we can end both the tragedy of the commons and the tragedy of conversion.

We may think it takes too much time to grow fences, but a living fence can mature to the state of being effective and productive in two years. It took nearly fifteen years to fully implement the Taylor Grazing Act in the United States.[25] After implementation, the western United States had millions of miles of barbed wire fences that provided one thing—control over the grazing of public lands. If we begin planting living fences now, in fifteen years we can have tens of thousands of living fences that protect and restore entire landscapes while providing animal feed, fruit and income to lift poor farmers out of extreme poverty. For farmers in the developing world, having a forest garden means that money really does grow on trees.

CHAPTER 4

HYDE'S SEEDS
JEKYLL'S SOIL

TRANSFORMATION OF SEEDS
LOSS OF SOIL

He who wants to feed the world
also wants to sell the seed.

—DAVE DEPPNER
Founder of Trees for the Future

SMOKE RINGS

A few years ago, a former Peace Corps Volunteer told me a story that made me realize the power of smoke. "For two-years I lived in a village in rural Togo," she began. "In the late afternoon, when the huts began to cast long shadows across the barren land, the villagers would collect brush and place it in small piles, just beyond the huts. As dusk fell, they would light the brush, making a ring of fires around the village. The smoke created a barrier of protection between the village and the evil spirits. No one was allowed to enter or leave the village until sunrise for fear of what lay beyond the smoke." This ancient practice most likely had its roots in protection from wild animals or warring tribes and is still practiced today out of habit and superstition. "I did not believe in evil spirits," she continued, "but with nightly practice and time, this tradition, and the fear it evoked, became truth and I really felt afraid."[1]

Like the unfounded fear evoked with this smoke, for decades, large agribusiness has been unleashing warnings of mass famine across the world, insisting that the world cannot feed itself without industrialized agriculture. By creating a smoke ring of fears, they have managed to hide the real evils that come in bags, bottles and seeds—the harmful effects of chemical fertilizers, herbicides, insecticides and genetically modified (GM) seeds. What we should fear is not what the future of agriculture holds without these additives, but what it holds if we continue using them.

Over the course of nearly a century, the notorious chemical company Monsanto introduced us to DDT, PCBs and Agent Orange, chemicals they claimed were harmless.[2,3] In the eighties, desperate for a new image and new products, Monsanto began experimenting with creating GM seeds. Alongside the science, they began to develop the strategy and tactics necessary to achieve industry dominance, in a world in which natural seeds would be virtually extinct.[3,4] As antiregulatory sentiment ran high in the U.S. in the early 90s, Monsanto found an open window for its strategy and the man to get them there, Michael Taylor. In 1991, after serving as Monsanto's attorney, Michael Taylor was appointed as the Deputy Director of Policy at the Food and Drug

Administration (FDA).[5] Despite concerns from experts at the Department of Health and Human Services[6,7] and some of the FDA's top scientists,[8,9] Taylor developed a policy that allowed GM foods to be used for both animal and human consumption, without FDA testing and without labeling. Under the Food, Drug and Cosmetic Act, amended in 1938, all food produced for sale in the United States must be tested and labeled if it contains altered substances or additives.[10] To circumvent this testing and labeling issue the FDA policy on GM foods states:

"The regulatory status of a food, irrespective of the method by which it is developed, is dependent upon objective characteristics of the food and the intended use of the food (or its components) . . . the key factors in reviewing safety concerns should be the characteristics of the food product, rather than the fact that the new methods are used."[11]

Or in simplified terms, if it looks like a duck and quacks like a duck, then the FDA will call it a duck, no matter how it was made. Once this policy was in place Taylor returned to Monsanto as their Vice President of Policy to implement the policy he'd designed.[12]

The problem is, GM food looks and acts like its genetic original on the surface, but it is the effects of the altered genetic information that we do not fully understand. Unlike tree grafting, selective breeding and hybridization that involve natural selection and cross breeding of *related species*, GM seeds (also known as transgenic seeds) involve the process of forcing genes from one species into another *entirely unrelated* species. The cells of living organisms have natural barriers to protect themselves against the introduction of DNA from a different species. Therefore, to introduce DNA from one organism into another, the new DNA must be attached to a bacteria (often E. coli) to infect the host cell or be shot into the host cell with a special gene gun. Because this is a bit like throwing a dart at a board, the genes do not always end up in the right place within the DNA sequence, thus thousands of trials and field tests are done to get the desired trait.[13,14]

Despite these challenges, Monsanto has been widely successful in the United States. Market dominance has been nearly achieved with GM corn (93 percent), GM soybeans (94 percent), and GM cotton (93 percent).[15] As we previously learned, 75 percent of all supermarket products contain corn in some form (as a starch, sugar, flour or oil) and nearly all of our meat, milk and eggs come from animals fed with GM corn or GM soybeans. Thus, with no U.S. mandates on labeling GM foods, is it becoming virtually impossible to avoid them. In a recent report the USDA stated, "Non-GM foods are available in the United States, but there is evidence that such foods represent a small share of retail food markets."[16]

The two most common GM seeds in the United States are Roundup Ready (herbicide tolerant) and Bt (insect resistant) products.[17,18] Roundup Ready seeds have been developed by inserting genes from a bacteria found to be immune to Roundup (Monsanto's toxic herbicide that kills all other plants). These seeds are engineered to be sprayed with massive amounts of Roundup and survive.[18] Likewise Bt seeds have been developed by inserting genes from a bacteria found to repel insects (a built-in insecticide).[19] However, the excessive use of the Roundup herbicide is creating "super-weeds"—there are currently fourteen weeds species that have become immune to Roundup. Several superbugs are also becoming immune to the Bt gene.[20]

Other Examples of GM Products Under Development

Strawberries injected with Artic fish genes to protect them from freezing.[21]

Goats injected with spider genes to produce milk with proteins stronger than kevlar for use in industrial and military products.[22]

Rabbits injected with jellyfish genes to make them glow in the dark for medical research.[23]

Enviro-Pigs injected with mouse genes to make their manure phosphorus free.[24]

Rice injected with human genes to produce pharmaceuticals.[25]

Proponents claim that GM foods are safe, and their short-term (ninety-day) in-house research shows no scientific evidence to prove otherwise.[26] However, a long-term, large-scale independent study on the effects of GM foods on mammals was conducted in 2012 by a team of scientists at France's Caen University, led by Professor Gilles-Eric Séralini.[27,28]

Following the protocol of Monsanto's previous 2004 study, Séralini and his colleagues reproduced the study over a longer period with more test subjects. Séralini 's study was conducted over a two-year period, using two-hundred rats. The GM test groups were fed a diet of 11 percent GM corn. The rats fed GM corn experienced reduced fertility, immune system failure, holes in their GI tract, organ systems failure, allergies and large tumors, particularly in the mammary glands. The length of the study was the crucial factor in determining the long-term health impacts, as the first tumors did not appear until four to seven months into the study. It was also determined that 80 percent of the rats exposed to low doses of Roundup developed up to three large tumors.[29, 30] Roundup, which is excessively sprayed on Roundup Ready GM crops, is made with carcinogenic and endocrine disrupting ingredients.[31] Thus regardless of the debate of over the safety of the GM seeds themselves, the necessary use of Roundup as a companion product is a serious public health problem alone.

When seeking out scientific research about the negative impacts of GM foods, we find that it is sparse. Many scholarly articles published in prestigious scientific journals, including Oxford's *Toxicological Sciences* journal, make statements similar to this,

> "The available scientific evidence indicates that the potential adverse health effects arising from biotechnology-derived foods are not different in nature from those created by conventional breeding practices for plant, animal, or microbial enhancement, and are already familiar to toxicologists. It is therefore important to recognize that the food product itself, rather than the process through which it is made, should be the focus of attention in assessing safety."[32]

If this sounds a lot like the FDA's statement, it's because it uses the FDA's policy and cross references other papers that used the same policy for the basis of their argument. The same phenomenon has occurred throughout Europe where GM foods were initially banned. Both before and after his tenure, as the chairman of the GMO Panel for the European Food Safety Authority (EFSA), Harry Kuiper worked for the International Life Sciences Institute (ILSI) task force led by a Monsanto employee and funded by some of largest GM seed producers. While working for the ILSI task force, Kuiper wrote several white papers that determined the requirements for risk assessment of GM plants and formed the basis for the comparative analysis concept that is still used by the EFSA today.[33]

"The underlying assumption of this comparative approach is that traditionally cultivated crops have a history of safe use for consumers and/or domesticated animals. These traditionally cultivated crops can thus serve as comparators when assessing the safety of GM plants and derived food and feed." (EFSA 2011)[34]

In short, the EFSA has adopted the FDA duck theory: if it looks like a duck and quacks like a duck, then let's just call it a duck, even if it was mixed with another species. So, we begin to see a trend. Just as the number of seed suppliers has consolidated, so has the number of expert analysts.[35] When digging deep we find that the research and regulations cited in pro-GM scholarly articles is either authored, funded, or both, by people involved in some way with large agribusiness or the biotech industry. This raises serious concerns over conflicts of interest and results in policymakers accepting the expansion of GM foods, the excessive use of chemical herbicides and insecticides, and the declining health of their citizens as unchallengeable realities.[36]

The seed giants claim that GM crops will "Heal, Fuel and Feed the world."[37] They aim to convince governments and wary farmers that GM crops are the sole adaptation strategy for ensuring agriculture productivity in the midst of population growth and climate change. They have even filed hundreds of multi-genome patents for what they are now calling "climate ready crops," claiming they have traits to support environmental stress tolerance such as drought, heat, flood, cold and salt.[38] It is easy to see why GM seeds appear to be the answer to our food production problems. They produce beautiful fields without weeds, they increase crop yields and reduce tillage. They decrease the amount of insecticide being sprayed (because an insecticidal gene exists inside the seed) and they are designed to survive in worsening environments.[39] Stacked varieties (with multiple genetic sources) even contain several traits making them resistant to herbicides and several types of insects.[40] However, the more farmers plant many of these seeds and spray the herbicides they require, the worse the environment gets, thus making it necessary to continue developing new and better seeds.

These seeds come at a high cost. After adjusting for inflation, the USDA found that the cost of GM seeds rose 50 percent between 2001 and 2010 and these seeds must be repurchased each year.[41] Although producers of GM seeds do not want labeling on the food their seeds produce, their bags of seeds are clearly labeled with a patent warning.

"The purchase of these seeds conveys no license under said patents to use these seeds or perform any of the methods covered by these patents. A license must first be obtained before these

seeds can be used in any way. See your seed dealer to sign a Monsanto Technology/Stewardship Agreement. Progeny of these seeds cannot be saved and used for planting or transferred to others for planting."[42]

This means that because these seeds are patented, they cannot be saved and reused or shared. They cannot even be planted without signing a license agreement. So farmers who have been saving and reusing seeds for centuries no longer have control over their seed supply. In the right conditions, corn pollen can travel hundreds of miles and still be fertile. Thus, controlling cross contamination between GM and non-GM fields is virtually impossible. Monsanto has a team of seventy-five investigators and lawyers to defend their products. Hundreds of farmers who have reused seeds or experienced GM contamination in their seed supply have been sued by Monsanto and lost for copyright infringement.[43]

Within two decades, armed with a formula, patents and a revolving door in regulatory agencies, Monsanto and nine other companies have gained control of 67 percent of the world's regulated seed supply.[44]

Seeds are not the only component of agriculture that is being altered by big agribusiness. Just as the soil must be stable for plant growth, it must also contain nutrients. With nutrient levels severely degraded in soils around the world, big agribusiness has engineered a solution for that too. The optimum nutrient conditions for plant growth involve three letters, N–P–K: nitrogen, phosphorus and potassium. These elements exist naturally in most soils, but intensive farming of monocrops requires more than the soil alone can provide. Likewise, if the soil is not replenished with these nutrients via organic material or manure, each year the soil becomes more deficient and yields less.[45] We know the land needs fertilizer; the debate lies in where this fertilizer should come from and its impact on people and their environment.

Nearly two hundred million tons of chemical fertilizers are applied worldwide every year.[46] Similar to claims about GM seeds, agribusiness claims that without chemical fertilizers the world could not feed itself. However, unlike natural methods of returning nutrients to the ground, chemical fertilizers are a band-aid. They are expensive, penetrate only the very top layer of the soil, run off with rains and are consumed by the crop with no lasting benefit to the soil the following year. Think of it like drinking a can of Red Bull: your body gets a quick fix of energy and then you crash, with no lasting benefits. Likewise, pouring chemical fertilizers on degraded soils gives them a quick fix, and then they crash with no lasting nutrients.

The most used phosphorus chemical fertilizer in the world is diammonium phosphate (DAP).[47] The extraction and processing of DAP is damaging to the environment and the health of the people who make it. It is then shipped via boats, trains, trucks and wagons to farms throughout Africa. The carbon and environmental footprint of DAP raises great concerns before it even gets to the field.[48] In Africa, farmers are encouraged to sprinkle DAP, which comes in a granular form, next to each maize plant, and miraculously the plant grows better.[49] However, nitrogen and phosphorus levels cannot be built up permanently by chemically fertilizing the land. When excess quantities of nitrogen are applied to a field, the nitrogen is not retained in the soil; it is lost by leaching and denitrification. Leaching of nitrogen into the groundwater or into streams is a serious form of water pollution.[50] Likewise, when plants are removed from a field, phosphorus goes with them, gradually reducing phosphorus over time. Farmers used to ensure that everything they took out of the ground was put back in, in the form of biomass (organic material). This created a closed-loop scenario, where phosphate would have the capacity to be reused up to forty-six times as food, fuel, fertilizer and food again. On the other hand, when phosphorus, in the form of DAP, is applied to nutrient-starved soil, it has the capacity to be used only once.[51] Excess amounts of DAP application run off into streams, creating algae blooms that pollute water and kill fish and other animals.

Some aid organizations are advocating for subsidization of chemical fertilizers and improved seeds for farmers in Africa. One such organization is The Alliance for a Green Revolution in Africa (AGRA). AGRA is a $400 million dollar enterprise initiated by the Rockefeller Foundation and the Bill & Melinda Gates Foundation.[52] AGRA claims to work for a transformational change in the areas of seeds, soils, market access, policy and partnerships to trigger a farmer-centered agricultural green revolution in Africa.[53]

This sounds amazing, until we remember that the first "green revolution" funded by the Rockefeller Foundation in Mexico and India destroyed local agriculture systems.[54] The top positions at AGRA and those of their grantees are filled with people from the agribusiness, biotech, seed and fertilizer industries.[55] The new vision and policy for agriculture aid in Africa has been laid out in several white papers funded by the Rockefeller Foundation and authored by the recently retired, former FDA Deputy Commissioner for Foods, once again, Michael Taylor.[56] With this knowledge, it seems that the kind of "transformational change" they are visioning is control over the seed supply, soil nutrients and herbicide for the continent of Africa, much like that which now exists throughout the United States.

As we have seen in Africa time and time again over the past fifty years, imported agriculture inputs do not "lift farmers up." They just shackle them to a dependency on expensive seeds, fertilizers, insecticides and herbicides that ultimately destroy their soils and environment, leaving them in worse condition than when they started. These imported products produce a new kind of indentureship, making rural farmers once again serfs to chemical and agribusiness companies of the same industrialized nations that they fought to free themselves from only decades ago. Although deforestation and desertification are slowly depleting the fertility of our soils, the solution cannot be engineered in a lab, nor can it be solved through a revolving door between industry, government and nonprofit governance.

There's got to be a better way, right?

There is.

LEAVES NO TRACE

When leaves fall in a forest, and there in no one to see them, do they still have an impact? They do on the soil. The leaves of trees and other organic plant material naturally replenish soil nutrients through plant-animal-soil-atmosphere interactions, creating different carbon neighborhoods of organic matter. We say neighborhoods because large communities of organisms live within healthy soils, tirelessly working to improve them. In just a handful of fertile soil there are more microorganisms than people that have ever lived on our planet.[57] In one teaspoon of healthy soil you can find "up to one billion bacteria, yards of fungal filaments, several thousand protozoa, and scores of nematodes."[58] Together these microorganisms bind soil together, help plant roots take up nutrients and fix nitrogen from the air for plants to use. Through the decomposition process, the soil is continuously restored by this microbial community.[59] These natural processes that build nutrients in the soil do not produce excess or create polluted runoff. However, when soils are sprayed with herbicides or insecticides, species essential to this process die and the soil loses its natural ability to regenerate. When soil and its associated organic matter is lost, the economic loss of the land and what it "could have" produced far exceeds the value associated with a bag containing N–P–K.[60] There is no replacement and no quick fix for this natural process. The only permanent solution comes with organic matter, minimal disruption to the soil and time. Thus, the costs associated with the loss of soil and its nutrients must also include the loss of time needed to restore it—silently and efficiently, below the surface, where the flora and fauna that produce nutrients leave no negative trace on the land, air or water.

In addition to the microscopic community's activities, there is another phenomenon taking place within the structure of the healthy soils that is fundamental to our understanding of how trees are the true stewards of our world. It involves trees' remarkable ability to communicate through nutrients. Dr.

Suzanne Simard, of the University of British Columbia's School of Forestry, has discovered that trees communicate with each other through nutrient sharing and chemical and electrical signals.[61] Like human mothers, mother trees (the older trees in the forest) can recognize their offspring and pass nutrients to baby trees when they sense deficiencies.[62] They can also share information. The root tips of trees have brain-like structures that allow them to pass chemical and electrical signals, much like human nerves. In healthy soils, the same fungal networks that help plants take up nutrients also help trees pass information throughout the forests, including warnings of insect attacks and fire. This allows other trees to activate their defense genes and close sap pours that can further fuel fires (this is just one instance among many). It has been found that this defensive communication message can be passed and activated in surrounding trees within six hours.[63] These recent discoveries of nature's wondrous operations show that the value of the microscopic and fungal communities in our soil is still being discovered. This value cannot be replaced with chemicals alone. The loss of these soil communities is creating a ripple effect of repercussions, inhibiting our ability to produce nutritious food and sustain biodiversity on our planet. As the land dies, the same fate awaits us.

GROWING NUTRIENTS

Families living in rural areas of developing countries have clear micronutrient deficiencies that can be met with better nutrition, but also with better soils. There is a direct relationship between micronutrient deficiencies in soils and in humans, particularly for iodine, selenium and zinc.[64] Since the first Green Revolution in the 60s, crop breeders have been developing hybrids of crop varieties that can grow crops in worsening environmental conditions. Although this has improved yields, it has decreased micronutrients. According to John Crawford, professor at the University of Sydney, "Modern wheat varieties have half the micronutrients of older strains, and it's pretty much the same for fruits and vegetables. The focus has been on breeding high-yield crops which can survive on degraded soil, so it's hardly surprising that 60 percent of the world's population is deficient in nutrients like iron. If it's not in the soil, it's not in our food."[65] We get our nutrients from the food we eat, and our food gets its nutrients from the soil. Currently, almost half of the nitrogen found in our bodies' muscle and organ tissue started out in a fertilizer factory.[66] If the soil is full of factory-built ingredients, genetically modified organisms and insecticides, our bodies will be as well. If the soil is healthy with natural fertilizer from organic matter and planted with unaltered seeds, the source of our nutrients will have come from the earth, as intended.

With the knowledge that there are both soil and human micro-nutrient deficiencies throughout Africa, agroforestry technicians can begin to address these needs by working with farmers to plant trees that will benefit both the soil and the family. This begins with Green Walls of trees that protect the fields from grazing animals and fierce winds. These green walls are full of what we call pioneer trees. Pioneer trees are fast-growing, nitrogen-fixing trees that grow well on degraded soils. They fix the nitrogen from the air for use in the soil, replenishing it in place. These trees are initially planted in rows around and across the cropping area, and dispersed throughout the site to protect the field, stop erosion and begin restoring fertility to degraded soils so that more valuable crops can be cultivated. Most of the pioneer species used can be frequently coppiced, meaning they grow back quickly after selective cutting. Therefore, they are constantly being cut back and allowed to regrow. The harvested branches can be used for timber or firewood, while the organic mounds of leaves and small leaflets are tilled into and mulched on top of the soil. While chemical fertilizers contain just a few macro nutrients (nitrogen, phosphorous, potassium), trees pull a variety of macro *and* micro-nutrients from the depths of the soil and recycle them into topsoil through their leaves. These leaves contain nutrients that are essential for people's health and which have long leached out of the soil—like iron, manganese, zinc and boron. By tilling leaves into the soil, farmers begin to restore the soil with minerals that help plants grow and ensure healthy nutrient-rich food.[67]

Farmers with forest gardens are finding that the best insecticides have wings. Monocultures usually attract swarms of a single insect that quickly infest a field. Use of insecticides kills all insects, even the beneficial ones that assist with pollination, including bees. Growing a variety of trees, plants and vegetables in a horizontal and vertical space creates a food web with a diverse population of insects that feed on each other and birds that feed on multiple insects. Some insects can survive on only one kind of crop. If that crop is grown every year, the population of the pest can build to very high levels.[68] However, intercropping of a variety of trees and vegetables can be used to repel insects, particularly when using plants with strong odors—such as garlic, basil, marigolds, rosemary, henna and onions—that confuse insects. Creating segmented vegetative barriers of different plant species can assist with crop rotation and also deter insects.[69] For example, suppose a caterpillar wants to eat tomatoes, but all he can smell are marigolds, and there is a thick, tall row of vetiver grass blocking his path. By the time he gets to the top of the grass he is exposed to a bird that swoops down and BAM, the food web naturally replaces chemical insecticide.

Africa is at an agricultural crossroads. Currently many African countries have a ban on GM seeds, and their use of chemical fertilizers is only 3 percent of the world's total.[70,71] For millennia, farmers have harvested and saved their seeds for replanting. However, with the patenting of GM seeds, they will no longer be allowed to do this. GM seeds are designed to work well only with other expensive inputs: chemical fertilizers and herbicides. Small farmers in Africa cannot afford these additional expenses, and their introduction in other developing countries failed to yield the miracles they claimed. Furthermore, I have seen herbicides and insecticides improperly used by farmers who cannot read, creating great health risks.

A decade ago many African countries went from no phone lines to cell phones, skipping a generation of invasive and resource heavy infrastructure. Some rural towns have gone from no electricity to solar and wind power, and some cities have gone from no trash collection to recycling. The lack of technological advances has, in some ways, put the continent in a position of choice. There is no doubt that agriculture production must increase in Africa to feed its growing population and prevent land grabs from outside. However, the lack of existing methodology, technology and regulations provides an enormous opportunity for African farmers to learn from the mistakes made by industrialized nations. They have an opportunity to choose a family-farm-centric approach that will protect their lands, increase production, ensure sustainability and create a growing livelihood for their families and communities. They can take back their sovereignty and reveal the smoke blowers for what they are: capitalists profiting off the backs and fields of the poor.

CHAPTER 5

TOO LITTLE
TOO MUCH
TOO DIRTY

WATER

*In an age when man has forgotten his origins
and is blind even to his most essential needs for
survival, water along with other resources has
become the victim of his indifference.*

—RACHEL CARSON

THE CRADLE WILL FALL

In 2012, Scott Harrison, the Founder and CEO of Charity: Water,
walked nine hours to the remote village of Meda in Tigray, the
northernmost region of Ethiopia. Once upon a time, Meda was
covered in forests and rich in water resources—water flowed year-
round. The forests are gone now, and like so many rural areas in

Africa, the town is dry and water is scarce. Scott made the arduous trip to discover the truth about a young girl named Letikiros Hailu, whose story seemed so tragic as to have all the makings of folklore. Upon learning more about Letikiros from her mother and other villagers and making the long walk for water with her friends, he found the truth both devastating and all too real.

Letikiros was only thirteen when she woke early one morning, tied a large ceramic pot to her back with rope and met her friends for the walk they made four days a week. She had made the trip hundreds of times since the age of eight. Following their initial long walk to the edge of a steep gorge, they carefully navigated the seven-hundred-foot decent to a place where the water seeped from the mountain, known as Arliew Spring. The spring dripped so slowly that it could only fill a few pots each hour. When they arrived, the line of women was already long. Realizing they wouldn't be able to fill their pots in time, Letikiros and her companions continued walking another three hours to the Bembya River, which always had plenty of water but was not clean. Descending another steep slope, they gathered water and helped each other tie the heavy pots to their backs. Then they began their long walk home. Not far from the village, just before dusk, Letikiros said goodbye to her friends, and they each went their separate ways. It is not certain what happened next, but it is believed that she must have stumbled and fallen. The pot broke and its precious contents that she had spent ten hours and all her energy collecting vanished into the dust. Left with only shards of a ceramic pot and a rope, Letikiros hung herself from a lone tree that still stood near her village—a village where forests, water and now hope had vanished.[1]

If Letikiros' death was an isolated suicide, we might be able to look the other way and find excuses for it other than the burden of lack of water access. But Leitikiros' death speaks to a fundamental question that is rising in every nation in the world. Is water a human right or a commodity? Access to water is inseparable from life. From the moment of conception we become fully developed humans by spending thirty-six weeks in water. Our bodies are made up of 55 to 60 percent water,[2] we die within a week without it and death is slow and painful after exposure to contaminants in it. Water is culturally significant in celebrating births, deaths, holidays and other religious ceremonies. It is used to heat, cool, brew, cook, cleanse and purify. Everywhere it is needed to grow food and sustain life. Thus, declaring water as a human right may seem superfluous. However, although we need it, access is not guaranteed. The natural weather cycle provides us with water in unpredictable ways: too much, too fast or too little. In addition, deforestation, over-pumping and climate change are all causing changes in groundwater tables and surface water availability. In Letikiros' village, the lack of water stems from Ethiopia's massive deforestation, where 98 percent of the forested area has been lost in the last fifty years.[3] The loss of this forest cover, in Ethiopia and areas throughout the world, means there are no longer tree roots to channel rainwater back into the ground and recharge springs and aquifers. Instead the rain water runs off the barren land taking fertile top-soils with it, forming massive gullies, rendering the land unproductive and unable grow food. However, nature is not the only factor in access to water. Every natural resource has become a commodity: land, trees, minerals, nutrients, seeds and, now, water.

Water rights, dams and privatization have all drawn borders around water resources, creating unfair distributions, granting access to those who pay and creating water businesses that are considered the new gold—blue gold. Yet water is transient and cyclical, moving through the air, over land, through our veins. Can you own something that is transient, that changes shape, that is so integral to ecosystems and to all life?[4]

Ownership of the most famous river in the world, the Nile, is in constant dispute. Historically, the Nile has been depicted as a river in Egypt; however, 85 percent of the flow from the Nile River originates in Ethiopia's upper and central highlands.[5] Despite the fact that Ethiopia's monsoon rains feed the Nile, arcane water rights treaties give Egypt the lion's share of its water use. Egypt's claims over the Nile became entrenched during British occupation. After Egypt gained its independence, Ethiopia argued that the British law was no longer valid. However, just six years after gaining independence Egypt audaciously signed an agreement with Sudan that gave them a 75:25 split respectively over Nile waters, leaving Ethiopia and seven other upriver Nile states with less than 1 percent.[6] To store the flow of this immense quantity of water, Egypt built the Aswan Dam in the middle of the desert, creating a reservoir that is 298 miles long and 10 miles wide, but only a few feet deep.[7] This water supply converted seven hundred thousand acres of traditionally flood irrigated land into perennial irrigated land and brought another eight hundred thousand acres into production, supplying Egypt with the irrigation water needed to grow food for its expanding population.[8,9] However, the nutrient-rich silt that was delivered with the annual floods now backs up behind the Aswan Dam, making Egyptian agriculture completely reliant on chemical fertilizer. In addition, the rapid evaporation of water on irrigated land has led to salinization and waterlogging. Furthermore, in the most water-stressed region in the world, the mismanagement of this water resource is hotly contested, as 12 percent of the Nile's annual flow evaporates from this shallow lake in the desert.[9] From this lake Nile water is pumped through hundreds of miles of canals to irrigate a sea of green monocultures of wheat, rice, cotton, flowers and vegetables. I have been to the American University of Cairo's Desert Development Center and seen the rows of potatoes that stretch to the horizon, grown for export. I couldn't help but think about the people who eat them with no idea of the environmental and human suffering these potatoes inflict.

Although Nile water has allowed Egyptian agriculture to grow, it is not prospering. Large scale monocrops have attracted devastating insect outbreaks. Alongside this agricultural growth, the population grew, tripling in the three decades following the completion of the dam.[10] However, Egypt is not the only country with a growing population in the region. Neighboring countries will no longer idly sit by while Egypt benefits from the Nile's waters. Ethiopia is as equally desperate to store and access this water. Each day, millions of precious gallons of water, needed to grow food for Ethiopia's growing population, evaporates from Lake Nasser in Egypt's desert.

Since the building of the Aswan Dam, Egypt has vowed to use all measures necessary to protect its water. To secure its status in the region and ensure the protection of the dam, Egypt signed a peace agreement with Israel, which also has an interest in Nile waters. Following the agreement, former Egyptian President Sadat claimed, "The only matter that could take Egypt to war again is water."[11] Historically, Ethiopia has been financially unable to build a dam on their own. Concerns over water conflict in the region and Egypt's international clout have allowed it to veto the World Bank financing that Ethiopia sought. In 2005, with one in eight Ethiopians in need of food aid, the former Ethiopian Prime Minister, Meles Zenawi stated, "While Egypt is taking the Nile water to transform the Sahara Desert into something green, we in Ethiopia are denied the possibility of using it to feed ourselves. If Egypt were to plan to stop Ethiopia from utilizing the Nile waters it would have to occupy Ethiopia and no country on Earth has done that in the past."[12]

Thus, in 2011 while Egypt was roiling in revolution in the aftermath of the Arab Spring, its government in shambles, Ethiopia made the bold decision to risk funding the dam project on its own, which equates to 60 percent of its annual budget.[13] The country has issued bonds to finance the project and Israel has purchased an unreported share of these bonds.[14] With the Ethiopian Grand Renaissance Dam nearly complete, and Israel as part owner of water that Ethiopia will take back from Egypt, there is growing tension in a region that has already reached a tipping point for conflict.

Although this dam will provide water and hydro power for Ethiopia, dams are rarely a solution for poor rural farmers, as it is certain that they will never have access to this water. A glimpse of who the dam will serve can already be seen. Over the last decade the Ethiopian government has leased seventeen million acres to foreign investors from countries including India, Saudi Arabia, Germany, the UK and others to grow food for export. This is equivalent to 38 percent of land currently utilized by small holder farmers.[15] Furthermore, contracts have been signed with neighboring countries to provide hydro power in exchange for foreign currency, while many poor rural Ethiopian farmers will continue to spend their nights in darkness.[16] During its construction, the Renaissance dam displaced twenty thousand rural farmers.[17] However, they are not alone. Over the last twenty years, the building of dams around the world has displaced more than forty million people, more than all the combined world conflicts and natural disasters.[18] The World Bank often serves as the lending institution and advisor for these dams, selling dams as a cure-all for water scarcity, irrigation needs and electrical generation. Banks, however, do not have the world's poor at heart. The World Bank needs to spend about $20 billion a year to make a return for their investors. Dams are a great way to spend a billion dollars in one place.[19] Or, as Peter Gleike, who has spent his life researching and advocating for decentralized water systems, so eloquently put it, "The World Bank knows how to spend a billion dollars in one place. What it doesn't know is how to do is spend a thousand dollars in a million places."[20] But this is what rural farmers in Africa and around the world need. They need small investments in sustainable solutions in millions of places that can work independent of large dams and turbines, pipelines and international investors.

Here in the U.S., we do not have to look across the world for water conflict, we need only to look west, where California is the Egypt of the United States. Although 88 percent of the flow of the Colorado River originates from rain and snowmelt in Colorado, Wyoming, Utah and New Mexico, at least 60 percent of the lower basin is allocated to southern California.[21]

Access to water from the Colorado River has allowed farmers in arid southern California to produce a large portion of the fresh fruits and vegetables consumed by Americans. However, because the Colorado River has not kept up with Southern California's growth, the state has constructed several large-scale water transfer projects that pump water from Northern California's Sacramento Delta to its more populated south. The largest of these projects is the State Water Project, which consist of seven hundred miles of pipelines and canals and pumps water up a two-thousand-foot elevation, over the Tehachapi Mountains.[22] This water access has also allowed Southern California to grow beyond its capacity, providing water it doesn't have to industries beyond agriculture. Although farms that raise crops account for 80 percent of the state's water use, there is another kind of rapidly growing farm—data farms.[23] Data farms store all those websites and Facebook posts, YouTube uploads and ever-increasing clouds of information. California has more than eight hundred data farms that use over 158 Olympic size swimming pools' worth of cooling water each day.[24] That is equivalent to the water needed to produce more than one billion peaches or two billion tomatoes.[25,26] Power plants also use millions of gallons of cooling water each day, and so does the refining of gasoline, the production of microchips and consumption in the households of forty million Californians.[27] With all these competing interests, California is getting very thirsty. Long periods of drought are compounding this problem, reducing the moisture in trees which in turn has led to enormous forest fires that have burned more than two million acres of trees in the last decade.[28] The loss of these trees complicates the problem even more by preventing the channeling of rain water back into the ground. Therefore, when long droughts are followed by unforgiving downpours, the rainwater runs off the barren, cracked, treeless land, flooding cities, creating mudslides, destroying property and failing to replenish groundwater tables or return moisture into the soils.

Both cradles of agriculture, Ethiopia and California have historically been blessed with rich soils and abundant water. However, they are each diminishing these resources at an alarming rate. Without soil and water, we lose our their ability to grow food. The high-cost modern agriculture solutions that have been offered to farmers are not working. Ethiopia, at the time of this writing, is in a government-mandated state of emergency due in part to conflict spurred from the loss of these resources.[29] In its fifth year of drought, California's water restrictions have left some farmers with zero water allocations for irrigation.[30] With changing rainfall patterns, water depletion, destruction of forests, mudslides and topsoil erosion, these cradles and the food they produce are starting to fall. The fall of these cradles and their surrounding branches is seen in the desperate cry of rural farming communities. Like Letikiros in Ethiopia, farmers are turning to suicide as the only way out of drought, high debt and hopelessness. In India over the last two decades, over three hundred thousand farmers took their own lives by hanging or ingesting the herbicide Roundup.[31] In Australia, a farmer commits suicide every four days, and in the United States, the occupation of farming has the highest suicide rate in the country.[32] The land has been crying for help for decades. Now there is a cry for help from the farmers who work the land. The farmers who toil year-round to bring us food. The farmers who, under the current unbalanced system, will succeed only if they find a sustainable way to grow water.

PANDORA'S WELL AND DINOSAUR PEE

People must have water for every aspect of life, but how it is accessed and managed is extremely complicated. Just as surface water (river) allocations are constantly in dispute, access to ground water is equally troublesome, often creating conflicts between farmers and herders, and neighbors and industries. Overuse of groundwater can also degrade the surrounding environment and its natural ability to regenerate. Working and traveling throughout Africa I have seen a frequent phenomenon that I have come to think of as Pandora's well. The first time I recognized this problem was in Gueye Cisse's village in Ngodiba, Senegal. For years, the village of Ngodiba had only a hand-dug, rock-lined well. Women pulled water from the well for drinking, washing clothes and watering their goats. About ten years before I arrived in the village, an NGO drilled a well and installed a pay pump that filled a water tower. Suddenly the village had central access to running water. With this new water source, villagers began to create gardens. At the same time, the water source attracted herders who came to the village to water their grazing animals. This created the age-old "farmers versus herders" battle. Crops and animals in close proximity require fences, so people began to cut nearby trees to build fences to protect their new gardens. The fences never lasted long because of the termites. Therefore, each year more trees were cut to repair or replace the fences. With increased animal traffic, vegetation became scarce and the ground surface compacted. New seedlings that popped up were either eaten or trampled by the herds that came to drink in the village. Thus, in very little time, trees became sparse and the land got browner and drier. When I arrived in the village the trees had almost disappeared. By climbing the water tower, to get a bird's eye view, I could see circles of deforestation that expanded out from the water source. Rather than being a springboard for life and sustainable development, the mismanagement of water in arid lands has created a growing ring of death around villages. Without sufficient training in sustainable land use, this cycle of water access and land degradation will continue to happen to the hundreds of thousands of water projects being implemented across the continent. Furthermore, the loss of tree roots inhibits the landscape's ability to recharge groundwater. And it all starts with the good intentions of Pandora's well.

Over-pumping from wells in developed countries can have different but equally devastating consequences. California, in its fifth year of extreme drought, has been pumping excessive amounts of ground water in its Central Valley region to make up for the water deficit. However, the aquifer is being severely over-pumped. Estimates reveal it would take fifty years of normal rainfall to recharge the aquifer if pumping ceased now.[33] However, it will never be able to fully recharge, because the ground above the aquifer is sinking. NASA satellites recently revealed that parts of Central Valley are sinking at a rate of two inches per year.[34] Over-pumping of an aquifer, without recharge, causes the clay-like soil to contract and the aquifer to collapse. The result of sinking land means massive loss of infrastructure. Already, water channels are buckling, sink holes are developing and housing foundations and roads are cracking[35] Once the land sinks into an aquifer, it cannot be recharged; thus, the most efficient, safest and purest way to store water is lost.

In the beginning, the earth was given one large ration of water and an amazing cycle that allows nature to cleanse it repeatedly. Over the Earth's 4.4 billion year history, water has not been created or destroyed, only transformed. The water we shower with today is the same water that witnessed evolution. The water we drink today was once dinosaur pee and Caesar's bath water.[36] Before Europeans came to America, it was a part of an intricate web of life that Native Americans understood. In 1854, the U.S. government sent a letter to Chief Seathl of the Suquamish tribe on the islands of the Puget Sound, off the coast of Seattle, requesting to buy their land. Soon after Chief Seathl wrote back with questions that we are still debating today.

"The President in Washington sends word that he wishes to buy our land. But how can you buy or sell the sky? the land? The idea is strange to us. If we do not own the freshness of the air and the sparkle of the water, how can you buy them? . . . The shining water that moves in the streams and rivers is not just water, but the blood of our ancestors. If we sell you our land, you must remember that it is sacred. Each glossy reflection in the clear waters of the lakes tells of events and memories in the life of my people. The water's murmur is the voice of my father's father. The rivers are our brothers. They quench our thirst. They carry our canoes and feed our children. So you must give the rivers the kindness that you would give any brother."[37]

When we look at the damaging transformation of our water resources, it is easy to say that we are ignoring this request in every sense. Rather than giving kindness to the river, as we would a brother, we have tortured our rivers in ways not fit for our enemies. Is water our enemy?

It appears so. Today we are polluting water in an irreversible way. However disagreeable it may seem to be drinking ancient dinosaur pee, neither the dinosaurs nor Caesar were capable of producing the 116,000 chemicals that enter water today in the form of pharmaceuticals, cosmetics, pesticides, herbicides and other products.[38] Unlike Pandora's well that creates a ring of deforestation around a village, the contamination of water does not remain centralized. The water cycle continues to evaporate and condense, dropping the contamination elsewhere. Likewise, the cosmetics that we use and chemicals and pharmaceuticals in our body end up in our toilet and shower water. This water makes its way to the wastewater treatment plant, where it is treated to take out biological contaminants but not low doses of chemical contaminants. The water is then returned to the river to mix with contaminants from agriculture (insecticides and fertilizer) and street runoff (oils, antifreeze, salt). In the river, marine life is exposed to all these chemicals. The water is then taken from the river and treated at potable water treatment plants for biological contaminants, but again not for most chemical contaminants. Again we drink, shower and flush our toilets with it, adding more chemicals, and so on. With each cycle, more chemical endocrine disruptors are added. The transformation that is occurring cannot be cleansed by nature, and water is no longer a silent witness. The rapid changes in marine animals is not evolution, but rather deformation as frogs and fish develop tumors, birth defects and complete loss of male reproductive organs.[39] In light of this contamination, many people have turned to bottled water, thinking that it is a safe solution. However, most bottled water comes from the same tap water or groundwater sources, with no higher level of treatment. In fact, companies like Nestle, who have seventy different bottled water brands including the well-known Deer Park, Poland Springs, Arrowhead and Ice Mountain are pumping from public groundwater sources in Michigan, California, Maine, Florida and others states, bottling it and selling back to the people.[40] In many places, Nestle acquires the land through a lease from the Department of Natural Resources or pays only residential water rates and receives millions of dollars in tax abatements. In Michigan, Nestle received a ninety-nine-year lease for less than two dollars a day. While Nestle makes $1.8 million a day by selling water that comes free out of the ground, our environment pays a high price in water stress as surrounding wells dry up, streams become mudflats and lake levels fall.[41] To Chief Seathl and his people, the buying and selling of natural resources was not a concept. Likewise, citizens near Nestle wells in Michigan, California and Florida never imagined a business could take control over public ground water, bottle it and sell it. However, whether it's a large business like Nestle or hundreds of individuals in an African village, overuse of groundwater resources, without regard for the elements that permit sustainable recharge, leads to Pandora's well and the exponential problems that ensue.

GROWING WATER: CHANNEL IT DOWNWARD

There is a great tragedy and irony in Ethiopia. Four million years ago, it supported such diverse life as to enable the evolution of modern man, and now it is on the brink of being unable to support life in any form.[42] Although water is the building block of life, it is also rapidly becoming the new oil: countries that have it will thrive and those that don't will go to war over it. Treating water like oil brought us to this point: pumping from ancient reserves, piping, channeling, storing and separating it from the land. We seek to control and reserve it for what and when we deem necessary. Often those who have the power to decide what and when, do not think much about the "who," the public—particularly the interest of poor rural farmers.

While a political battle is ensuing over Ethiopia's water rights, rural farmers have no control over this. The Renaissance dam will soon be completed, but most rural farmers will never see a drop of it. Drought-stricken countries are already buying up land near the dam to produce large fields of irrigated grains for export to their own hungry populations.[43] Therefore, rural Ethiopian farmers must focus on the only thing they have control over—the water that falls on their land.

Unlike other resources, water returns to us from the sky in unpredictable ways. While our projects in East Africa experience increasing droughts, our projects in Senegal see flooding during times of the year when rain never fell before. Managing these unpredictable rainfall patterns requires a fundamental change in the way farmers produce food on their land. Decades of abuse through monocropping, overgrazing and the application of agrochemicals have robbed the land of trees, soil, moisture and nutrients. Ethiopia's land now has the physical properties and nutrient equivalence of concrete. When rain falls on this concrete-like land, the water makes its way to Egypt to irrigate not only agriculture but twenty-one-hole golf courses and the swimming pools of wealthy nationals and foreigners.

Farmers can take back control of this water by planting a diverse combination of trees and crops on their land that absorb and infiltrate water, acting as a living sponge during the rainy season. This starts with the roots of thousands of trees planted in a Green Wall around the field, creating pathways for water to channel back into the ground. As the leaves of these trees fall they provide food for microscopic life in the soil. When the soil is revived with a thriving microscopic community, it can retain more water for plant use.[44] This water stored in the soil is called green water and can be used directly by plants, preventing the need for irrigation.[45] Mulching the land with tree leaves can help retain 20 to 70 percent of this green water for an entire dry season.[46] We see this in our projects. The rains come in August and September, but the trees hold and use this green water for seven months, delivering ripe, juicy fruits in April during the driest time of the year.

Planting forest gardens allows farmers to take tangible actions to manage water on their land before flooding or drought occurs. This becomes increasingly important as climate and weather patterns change. In a monocrop system, if the rains come too early or too late, the crop is lost. In a forest garden, whether the rains come in early June or late August, the trees act as a natural storage and release system. This ensures that the farmer is not dependent on large dams, irrigation cannels and water rights decisions made in air-conditioned high rises. The farmer has his or her own hidden water storage system. Farmers begin to have control again. And not only control over the water on their land, but control over what they grow. Trees give farmers the ability to put the limited amount of water they have to use throughout the year. In the dry season, when vegetable gardening ends, tree fruiting begins. When nothing else is growing, and the monocrop fields lie dormant, forest garden farmers are harvesting jujube berries and cashews fruits while they wait for the next cycle of annual cropping to begin. The trees and mulch shade the soils from excessive evaporation, keeping soil neighborhoods alive and ready to absorb the monsoon rains. And when the rains come, the tree roots hold the soil in place and channel the water back into the ground to be used again during the dry season, creating a microclimate of water cycles and preventing Ethiopia's water resources from evaporating in Egypt's desert.

Like industrial farming, large centralized water infrastructure will only accelerate our collapse. For small farmers, these systems are out of reach. As the future climate becomes more unpredictable, farmers must adopt a methodology that protects their land, diversifies their crops and optimizes their natural resources. From the clouds to the surface and below, trees bring the rain and capture the water that falls. Forest gardens allow farmers, at no cost, to plant rain and grow water.

CHAPTER 6

WRITTEN IN STONE

RESOURCE DEPLETION/
BIODIVERSITY LOSS

*The Rabbit eating the berry, thanks the tree, and
the bird who knocked it down.*

—WEST AFRICAN PROVERB

ANTHROPOCENE: MAN CONTROLS THE
EXISTENCE OF ALL LIVING THINGS

If asked what the largest living organism is, we may stop a minute and think, "hmmm, a whale." Few people ever guess *a tree*. With a root structure below the ground as large as their structure above ground, trees are the largest living organisms.[1] Yet, we have been able to destroy a great deal of them with very little effort. Eighty percent of the world's natural forests have been destroyed.[2] Why? Because while trees are large, they are also immensely useful. Trees provide us with houses and firewood, furniture and telephone poles, paper and pencils, toilet paper and coffee stirrers. Who could live without those?

In the U.S., we were blessed with an abundance of large, tall, straight trees that we used to develop our nation; to build ships and ports, to shore foundations and erect houses, to construct water wheels and carriage wheels, to fuel factories and install thousands of miles of railroad tracks and telephone and electrical lines.[3] Early on in this fury of development, the United States destroyed 90 percent of our native forests.[4] Not far from where I live, the old Cass Railroad in West Virginia is a haunting example of how good we were at systematically removing America's tree cover. We used trees to build railroads to accelerate the rate at which we could clear tree cover across entire mountain ranges and get the timber to market. In Central Maryland, I take my family camping around Catoctin Iron Furnace Park where a wood-fueled iron furnace wiped out massive swaths of old-growth forests. Secondary forests did regrow in this area and are now protected. In many areas of the U.S. we are in our second and third rotation of mining unprotected forests (particularly in the South and Northwest). However, our nation was fortunate to have early conservationists, including writer and artist John Muir and President Theodore Roosevelt, who brought attention to our forests and made conservation a priority.

Most developing nations that once had an abundance of trees did not have the opportunity to use them to develop. Instead, colonizing nations, whose own forests had already been largely exploited, clear cut and exported their forests. Such is the case with one of the most tragically impoverished countries in the world: Haiti. Like its neighbor, the Dominican Republic, Haiti was once covered with trees. During nearly two centuries of colonization, the French brought tens of thousands of slaves from west and central Africa to the island. For the return voyage they filled their ships with trees.[5] On clear-cut Haitian land, they began to develop plantations of monocrops, primarily coffee, tobacco, indigo and sugarcane. With free timber exports, slave labor and massive cash-crop production, Haiti became France's most profitable colony.[5,6] In 1804, after years of uprising, the slaves defeated a twenty-thousand-man French army. However unbelievable it may seem, the French demanded payment for the loss of their slaves and colonial plantations, and slapped a large indemnity fine on Haiti.[6] For nearly a century, the Haitian government sold and exported trees and continued intensive cash cropping to pay for their independence from France. These activities decimated much of the remaining forests. The remaining 30 percent of forests disappeared slowly as wood became the poverty-stricken country's primary fuel source for cooking and heating.[7,8] It is the only country I've seen where citrus and breadfruit trees are being cut and sold for firewood. As Haiti's trees disappeared, the mountainous island became susceptible to tropical storms and hurricanes. Without trees and roots to hold the soil in place, Haiti's soil has flowed into the ocean. Many Haitians now say, "The mountains are getting old, we can see their bones beneath their skin."[9] When only the land's bones are left, it becomes nearly impossible to reforest the land and grow food. For Haiti, conservation efforts may have come too late; it is now facing a mass extinction crisis, particularly of frogs,[10] which are Haiti's primary mechanism for insect control.

By understanding the context of Haiti's biodiversity loss, we can prevent and reverse it in other areas. Cameroon, the country of origin for many Haitians,[11] is falling victim to modern-day colonization. In 2009, a subsidiary of U.S. agribusiness firm Herakles Farms sat down with a map and carved out 150,000 acres of Cameroon's unprotected forest, the size of Singapore.[12] The company was given a ninety-nine year lease on the land to develop a palm oil plantation at a cost of less than fifty cents per acre per year.[12,13] Although this area is not yet protected, it is not degraded land and it is not an empty forest.[14] It is located in one of the most biodiverse areas in the world, surrounded on all sides by national parks, wildlife sanctuaries, forest reserves and proposed reserves.[15] Palm oil plantations have already destroyed much of Malaysian and Indonesian rainforests, where 90 percent of its production occurs.[16] To develop palm oil plantations, the forest is clear cut and burned. This not only displaces thousands of animals and birds that can escape, but kills millions of smaller creatures that can't. It also destroys forest resources that thousands of people depended on for food and livelihoods. Do not be fooled, green does not equal green—a plantation of palm trees is not forest.

We may not think that losing trees in developing countries can directly impact us, but it does in numerous ways. Seventy percent of the world's land creatures depend on forest for their survival.[17] With the loss of trees in the southern hemisphere we lose migratory birds. Migratory birds cannot just stop migrating; they die. With the loss of trees, we lose all the organisms that depend on them—microorganisms, bacteria, plants, insects, amphibians, reptiles, birds and mammals. By the most conservative estimate we are losing about twenty-four species per day.[18] The week is flying by, literally—look at the calendar. Today, twenty-four species will become extinct. Tomorrow twenty-four more, over the weekend forty-eight more. How many species will be lost by your next birthday? Labor Day? The New Year? The ecological web is so complex, our understanding has only scratched the surface of the immense changes that will occur with each loss of species. Haiti's butterflies can now only be seen in paintings that hang in hotels, and Kenyan farmers don't recall seeing a butterfly in years. A third of frog species are critically endangered.[19] Bees and bats are getting wiped out globally. One in eight birds (the superheroes of all species) are threatened, and their numbers are declining rapidly.[20] These creatures are our canaries in the coal mine. They are also our pollinators and insect eaters. Without them, fruits and vegetables stop growing and mosquitoes proliferate.

Rapid loss of plant and animal species, land use changes and measured levels of pollutants in sediments have led geologists to consider that we have entered a new geological epoch, the Anthropocene, where one species—humans—is the dominant influence on the environment.[21] Through the study of rock formation over time, geologists use epochs to signify massive changes in rock series due to climatic and environmental conditions that affect life on earth. Changes in rock series are nature's history book, telling the unwritten story of earth's long past.[22] Thirty-two epochs have been defined, spanning from the beginning of life on earth to the present. We are very familiar with some epochs, such as the early, middle and late Jurassic, when the last dinosaurs existed.[23] For the last 11,700 years we have been in the Holocene epoch, which marks the retreat of the last major ice age. In 2012, scientist proposed that the massive changes occurring on earth indicate we are entering a new epoch, the Anthropocene—the epoch of Man. Several proposals have been sent to the International Commission on Stratigraphy (the governing body that sets boundaries on geologic ages) with regard to the start date of the Anthropocene.[24] However, they all agree on one thing: the changes man has, and is, making on earth will signify a new strata in the rock series, an undeniable marker for life in future epochs to read.[25] Or, in other words: rock doesn't lie. It is frightening to think our geologic story has already begun and there is no way to erase or edit what has already been written.

When deciding on a start date of this new epoch, evidence must be considered that could be detected worldwide. Some are proposing a start near 1950 when radioactive material dispersed throughout the globe during the use and testing of the atomic bomb. This was also a time that fly ash began being detected in lake sediments around the world, signifying the beginning of coal burning.[26] The large oil-drilling operations that extract million-year-old oil from previous rock strata, fracking for gas, mining that removes entire mountaintops and and deforestation that changes land from forest to agriculture, can all be detected in the formation of sediments that become rock. But perhaps what will be most telling are fossil remains.[27] Mass graves of animal bones, similar to those of dinosaurs and mastodons, can be found all over the planet, not due to a meteor hitting the earth, an ice age or volcanic activity. The mass graves of beavers, otters, lynxes, sables, foxes and martens found through the western United States, Russia and Siberia point directly to man and the fur trade.[28] Bones of thousands of elephants in Africa with missing tusks point to man and the ivory trade. Today we want to believe that our thinking has evolved, that we no longer condone the mass killing of animals, and to some extent it is true. We have learned to value the mammals of the world, but we have not learned to value their environment and the lifeforms they depend on. Who cares if we save the otters or whales or elephants, but destroy their food source? We need to ask ourselves, which is more humane: a slow death by starvation, or a ruthless slaughter? This relatively short but tragic story is already written in our sediments. And it continues in new ways.

Not only are we losing biodiversity, we are losing agrobiodiversity. Over the last few hundred years we have committed a new sort of geocide on nature, through human-selection. In the millions of years prior to man's existence, nature selected what would survive and what wouldn't. Darwin spent his life researching and defining the theory behind this process as "natural selection." In a relatively short time we have interfered with this process by selecting a handful of trees, grains, vegetables and animals that we deem vitally important and breeding them at the expense of all other species. In each of the five former mass extinctions, it was nature that created them.[29] Now it is humans who are deciding what thrives and what doesn't. Humans have selected palm oil, fruit trees, cacao, coffee, wheat, corn, soy, rice, sugar, about a dozen vegetables, cattle, pigs, chickens and a few fish, to be served up as the food staples around the world. Seventy-five percent of the world's food production comes from just twelve plants and five animal species.[30] Sixty percent of all human calorie intake comes from only three plants: wheat, corn and rice. Imagine that: three out of the estimated thirty-thousand known edible plant species.[31] The ratio is astonishing and borders on insanity. In the last hundred years, three quarters of plant diversity has been lost as farmers traded local varieties for genetically uniform high-yielding varieties.[32] Now that GM crops have been added to the mix, there will be an increased reliance on vast monocultures of species that we have created.[33] In addition, large fishing operations and bottom trawlers that indiscriminately catch everything destroy habitat and throw away the less marketable catch are depleting the world's oceans.[34] Likewise, wetlands and mangroves are being destroyed for large-scale aquaculture of a few species including shrimp, tilapia and salmon. Mangroves serve as fish nurseries and are essential to the natural restocking of our dwindling fish populations. The mass cultivation and breeding of select species in monocrop fields, plantations, fish farms and confined animal feeding operations (CAFOs) have led to the loss of other species through destruction of habitat, land degradation and water pollution.

Although geologists have officially announced the beginning of the Anthropocene epoch, it is estimated that it will take three years to determine a date for when it began.[35] The more pressing question is, will the beginning of a new epoch wake us up, or will it be another headline? Oceanographer Sylvia Earle, who has been studying oceans since the year we landed on the moon, sounds the alarm by stating, "The next ten years will be the most important in the next ten thousand."[36] What story will the geologists of the future read in the rock of their ancestors? A story of all that we destroyed or a story of all that we saved?

WE WANT MORE! WE WANT ENOUGH!

As humans, once our basic needs are fulfilled, we have a tendency to want and use more. More food, more room in our house, more clothes, more cars, more land, more water, more energy. However, the earth's population has reached a tipping point. Speaking in terms of resources, it is impossible for us all to have more. If the entire world rose to the same consumption rates as Americans, it would be as if the population grew from seven billion to seventy-two billion overnight.[37] This scenario is silently playing out. The growing middle class in China, India and other countries is rising and wanting more,[38] yet the earth's resources are limited, water is limited and land is limited. Among the race for more, there is another half of the world's population that just wants enough: enough shelter to keep the sun and rain off their heads, enough nutritious food to eat, enough water to drink and cook and bathe, enough energy to see at night, enough money to send their kids to school.

In an effort to have enough, many rural poor turn to the forest for resources. In Western countries, many of us have a limited view of the use of trees and forests, citing wood, fruit, nuts and shade as their primary values. However, in rural areas of Cameroon, Niger, Senegal, Ethiopia, Kenya, the Congo and other African nations, people have not forgotten the immense value of the forests and the products they can harvest, sell and use from them. In addition to using trees for building and firewood, people rely heavily on the forest for other resources, often referred to as non-timber forest products (NTFP).[39] NTFPs include bark, roots, tubers, corms, leaves, flowers, seeds, fruits, sap, resins, oil, fungi and animal products—all harvested from natural forests.[40] NTFPs are used for food and making handicrafts. Seventy percent of the world's medicinal needs are derived from NTFPs.[41] They are also important sources of income for rural residents. It is estimated that as many as 1.6 billion (roughly a quarter) of the world's poorest people rely, in some way, on forests for their livelihoods.[42]

Living at the margins of society, subsistence farmers are largely unaffected by economic and political systems. The factors that affect their survival are weather, loss of livestock, crop failure and illness or death of a family member. During these times of crisis their primary coping mechanisms involve selling forest products.[43] In times of need, the forest become a security net, giving people what they need to survive, giving them enough.

The southwest region of Cameroon is just one example of competing interests to globally have more and locally have enough. This region of Cameroon is one of the most biologically diverse habitats in Africa, holding some of the highest levels of flora and fauna densities on the continent with over 9,000 plant species, 550 fresh water fish species and 1,200 bird and animal species (of which 630 are critically endangered and threatened), including the Nigeria-Cameroon chimpanzee and the Cross River gorilla.[44,45] Living in these communities are also some of the poorest people in the world who depend on agriculture and the forest for their survival. In Cameroon, up to 46 percent of remote rural household income comes from the forests.[46] There are numerous indigenous vegetables, spices and medicinal plants that are highly sought after and over-exploited in the forest for food and income, including bush pepper and bush onion (used as spices and condiments), njangsa and bush mango (used as thickening agents for sauce), safau (a West African plum) and others. In addition to rural residents mining forest products, there are large companies that search for specific forest resources, like the highly-sought-after African cherry bark (which has been overexploited by pharmaceutical companies for its use to treat prostate problems). Alongside these competing interests are large plantation developers who are clear cutting areas to plant the cash crops of cocoa and palm oil for export to places that "want more." Like corn and soy, palm oil has become a common ingredient in our everyday products (estimated to be in 50 percent of all supermarket products)[47] and ever present in processed and fast food. In the rush to have more and to have enough, natural forests and the species they support are losing ground.

In many areas, the combination of deforestation, growing populations and need for farmland or plantation land has left islands of native forests among a sea of agriculture. Yet 70 percent of the world's animals and plants depend on these forests for their food and habitat.[48] In order for these islands of species to survive, they must share borders with ecological farming practices. Ecologists have found that the makeup of this sea of agriculture is as important to the survival of species as the conservation of the very island forests in which they live.[49] Although conservationists and agriculture developers have historically been at odds over land use, the need for an ecological border between food production and forests calls for these unlikely allies to unite.[50]

Together conservationists and practitioners of sustainable ecological agriculture can combine their knowledge to: eliminate pressures to convert forest to cropland, provide corridors that connect island forests and find ways for the forest to provide for agriculture needs, i.e., natural fertilizers and pest eaters.

GROWING MORE

There is evidence that we no longer want the responsibility of controlling the existence of all living things.[51] Publicly, the human-versus-nature debate has evolved from a "man controls nature" narrative to a "we live together" philosophy. However, for several reasons, it's too late. We can't walk away and wash our hands of the disaster we have created. Now the only way to reverse the environmental damage resulting from human intervention is with, however redundant this may sound, more human intervention. But not just any intervention. We are out of time for experiments.

For many developing countries, which have already lost much of their natural resources, the only way to have more is to *grow more* in the most diverse and efficient manner possible, like the forest does. This requires converting traditional degraded farm lands into forest gardens. But how do you build a forest? In the age of the Anthropocene, you start with making man a part of it, not with the conventional goal of planting forests for the sake of conservation, but with a new purpose: planting forest for sake of food and fodder, for biodiversity and wood, for water protection, soil stability and for all the things that sustain life in all its forms. As David Jacke, author of *Edible Gardens* explains, "Forest gardening helps us take our rightful place as part of nature doing nature's work, rather than as separate entities intervening in and dominating the natural world."[52] By developing forest gardens, people can grow what they would harvest from natural forests, create corridors between agriculture and natural forests and revive biodiversity. With intensive monocrop farming, farmers toil all year using fertilizer, expensive seeds and insecticide; spending money and time to let one species survive at the expense of all others. Forest gardens reduce these efforts and costs by letting nature do what it wants to do. In many rural farming areas, if the field was left alone for many years, it would return to forest.

As a botanist and leading agroforestry technician, Limbi Blessing works for our partnering organization the Environmental and Rural Development Foundation (ERuDeF) in the southwest region of Cameroon. Dedicated not only to growing enough, but growing more, Limbi works with communities in the Mount Cameroon area to ensure a better life for farmers and preserve her country's beautiful land. Daily, she faces the great challenge of reducing the pressure on the natural environment by helping people grow what they need from the forest. While an agronomist often focuses on how to solve the farmer's problem of increasing productivity of one or two species through better techniques, the agroforestry technician focuses on helping farmers determine what combination of crops they need to grow to sustain themselves and their land throughout the year.[53] This is a fundamental difference in philosophy. The first focuses on how the "sick farm" can be fixed. The second creates a vision for how the farm and the family can live in balance.[54] By developing forest gardens that mimic the real forest, incorporating fruit, vegetables and indigenous tree species, Limbi and her master farmers are helping over seven thousand people produce food, wood and the other forest products they need, improving the lives of farmers and relieving the pressure on this bio-diverse ecosystem.

Integrating trees and other forest species into agriculture not only provides forest products close to the farmer's home, but also reduces the drivers for forest exploitation that come from one-crop farming systems. In addition to plants, other popular forest species can be integrated into forest gardens, including snails, bees and other sought-after species that create eco-webs and food chains, which in turn support natural environments and create new sources of income. Snails dine on leaves of *Leucaena* and *Calliandra* tree species and are coveted regionally, providing an excellent income at no cost to the forest or the family. Small-scale bee keeping provides an abundance of pollinators while producing income and a food source through honey, again with no harm to the forest. Forest gardens can also be designed to sustainable feed rabbits, poultry and livestock.

In many tropical areas including Ghana, Nigeria and Cameroon, farmers depend on and have a close relationship with the forests that remain. In Cameroon, the forest bush mango ripens at a time when there is no agriculture income. Families use this seasonal downtime in agriculture to harvest and process this forest fruit.[55] They receive half of their cash income during this seasonal forest harvest.[56] Here we see an opportunity to introduce bush mango into the families' forest gardens. In Ghana, harvesting of rattan from the forest sharply increases just before the seasonal cocoa harvest. Rattan is used to weave the baskets needed to harvest cocoa beans.[57] Again rattan can be grown in the forest garden, producing a sustainable supply close to home and eliminating the need to harvest from the forest.

In Mali, West Africa, where 90 percent of energy use comes from burning wood, people are depleting what French colonialists did not take—the remaining trees, for charcoal.[58] The demand for firewood for 1.8 million people in Bamako, the fastest growing city in Africa,[59] creates an endless flow of cars, pickups, minibuses and tractor trailers filled with wood and charcoal. Three decades ago, Suleiman Diarre, director of Forestry Management and Reforestry in Mali told the *New York Times*, "When I was a boy in the 1960s, people went three to ten miles outside of Bamako to get firewood. Today, they go one hundred miles out of town to find wood."[60] Since then, the population and demand for wood has grown. During a trip to Mali, I witnessed a line of vehicles stretching over two hours to the source. When I followed the line of vehicles a couple hours south of Bamako, I found groups of women burning brush beneath trees in a pocket of forest that took between several decades to hundreds of years to grow. The undergrowth was burned to remove prickly vines and chase away snakes and other venomous creatures so they could easily harvest all the remaining timber. In contrast, families who plant forest gardens often become self-sufficient in fuelwood by simply collecting the massive amounts of wood pruned from their trees. By growing a sustainable source of firewood, farmers not only supply their own energy needs but also sell excess for income.

The fruits, vines, leaves, wood and other products that farmers harvest from the forest provide necessities and much-needed income for monocrop farmers. But at what cost? By over-harvesting these forest resources, a farmer's income becomes a debt for the many species that depend on those fruits and habitat to survive. By asking the right questions, agroforestry technicians can help farmers design forest gardens that incorporate species of trees, plants and vines that produce the most sought-after forest products, growing a diverse array of crops to eat, sell and use throughout the year. This ensures that farmers have an abundance of these products without harvesting them from local forests and that the forest species and the species that depend on them have enough to continue to thrive.

Furthermore, as the competition to "have more" grows, it becomes ever more important for rural farmers to show that their lands are *growing more*. It is much harder for a government to give away thousands of acres of productive, income-generating, food-producing land than it is for them to give away thousands of acres of denuded, degraded land. By developing sustainable forest gardens that produce more on small plots of land, farmers add value to their land and value gives power. In the last one hundred years we have come to believe that controlling what the land grows and doesn't grow is powerful. But the real power lies in the wealth of biodiversity that can exist on a small plot of land. Darwin has shown us that we will need this biodiversity to adapt in the future. It is the only thing that will save us from ourselves and ensure that the age of Anthropocene does not end with the extinction of man.

CHAPTER 7

THE HUNGER BAR

HUNGER, POVERTY
AND HEALTH

*When I give food to the poor, they call me a
saint. When I ask why the poor have no food,
they call me a communist.*

—DOM HELDER CAMARA
Brazilian Roman Catholic
Archbishop

RATS!

"Hurry, you have to find something to eat! Hurry, you're losing energy!" my son yells, while playing the online survival game *Minecraft*. In the game, players learn to use natural resources to build and survive. Not unlike the real world, they use trees to craft houses, fences and make charcoal. They mine gold for technological uses, and redstone to create electrical circuits. They hunt and breed animals and grow plants to eat. The only hunger my son has ever known is the hunger bar in *Minecraft*. If he travels far from his food source and doesn't find something to eat before his hunger bar gets low, he will lose his ability to run and find or produce food. His energy level will slowly drop until he gets down to one-twentieth of his life. At that point he doesn't die, but he can no longer move. This is remarkably similar to real hunger.

Although my son has never experienced hunger, I have. During my first year in Senegal, just before the rainy season, I was famished. I had lost twelve pounds and become a vegetarian by default. To be honest, I hadn't actually seen a vegetable in months, I was more like a *carbotarian*. Gueye Cisse, my host mother, would put a small leaf or a flower on top of our rice at lunch and millet at dinner, trying to garnish the food and make it look enticing. It didn't work. Although I was hungry, it was difficult to eat because afterward my body was still hungry, craving nutrients. Some days I felt so weak in the afternoon heat that even swatting the flies away from my face took too much work. One late afternoon, as the heat rose from the baked earth and created a haze that almost looked like fog, my friend Omar, his cousin and I trudged out to their peanut field to begin preparing it for planting. Just before the sun set, Omar got lucky and caught a rat. This was not a street rat or sewer rat; it was a large field rat, about the size of a small cat. At first it didn't occur to me why they were so excited, but soon I understood. We took the rat back to my hut and skinned it, gutted it and began stewing it over my small gas burner. For flavoring we added the only other ingredients we had: salt and pepper. The sun had set and it was dark in the hut. The three of us just sat there by candlelight, mouths watering with the scent of cooking meat. When we were certain it was thoroughly cooked, Omar placed it on a plate and we each took pieces, savoring the small amounts of meat, cleaning the bones. When there was only one piece left, Omar pushed it toward me and nodded, "you take the last piece." I did. As I gnawed on it, it seemed bonier than the other pieces, and to have strange, flat . . . teeth.

"Ahhhh!" I screamed, throwing it several feet in the air. The candle went out. The hut echoed with Omar and his cousin's laughter as I searched around in the dark for water to rinse my mouth.

When we finally got the candle lit again, Omar put this hand on my shoulder and tried to speak between fits of laughter.

"I noticed you don't eat the heads of fish or chicken. I wanted to see if you were hungry enough to eat the head of a rat."

I think often about the hunger I felt in Senegal. Once I experienced it, I feared it, and never again questioned how far down the food chain I would go when faced with hunger. Though rats are not endangered, some of the other things I ate could have been, including monitor lizards and exotic birds. Other volunteers ate monkeys, snakes and wild boars, giving us just a glimpse into how hunger is decimating biodiversity.

In Senegal, most communities experience hunger cyclically throughout their lives. I say communities, because hunger is communal in Africa. When one family or a couple families have no food, others share what little they have. Although hunger is never talked about directly, with every greeting, language revolves around hunger in the form of questions about the health of the family. Hunger and health are directly related, and the statistics are numbing. Feel free to look them up. I can convey that a child dies every ten seconds from malnutrition-related illnesses,[1] but telling hunger's story in numbers desensitizes us to its debilitating reality.

Suffering from hunger is vastly different than feeling hungry. At the end of the day when my stomach is growling and I feel weak because I skipped lunch, I can grab a granola bar or a banana or some other quick energy food. When I get home from work, I can cook up something on the stove within twenty minutes. After I eat, I feel strong and vital again. However, in rural areas of the developing world, people cannot just buy a banana, much less a granola bar. Even if they had money, they couldn't find one. In these areas, obtaining food and cooking it is a long process, and the food staples that people can afford are limited in nutrients. It takes women half a day to make a simple lunch of boiled millet and milk. First, women wake up early to fetch water. Then they must walk miles to find firewood. Because they cannot afford to have their grain milled, they pound it themselves, which takes more time and energy. To cook, they light fires and then must wait for the coals to get hot enough. They put their kettles of water on their fires and continue to stoke them so that they stay hot. To get milk, women must milk their goat or cow. About five hours after they began this process, they serve the meal to their family. After eating, their stomachs are full, but their bodies feel weak. There are limited nutrients in boiled millet and milk, particularly when it's eaten day after day. Rather than creating the feeling of an empty stomach, hunger creates weakness, which leads to lack of energy to perform the great task of seeking out, growing and preparing food. But that was just one meal; these women must find the energy to put one foot in front of the other, to provide food again and again and again, to keep themselves and their families alive. But with each day the weakness grows. This is hunger. And when the weakness overtakes the ability to find and prepare food, this is famine.

From the middle of the dry season until the next harvest season (approximately five months) is the most difficult time. There is little fresh food available, food prices are high and the money from the previous year's harvest is running low. This is what people throughout Africa call *the lean time*. Coupled with increased hunger during the lean time is the extreme heat. Adults eat to survive, but children lose the desire to eat. Some women describe times when there is no food—nothing to give their children who are crying and inconsolable. They put a kettle of water on the fire and tell their children that the food will come soon to calm them down.[2] When there is food, the meals are all the same, families become quiet, mothers become heartbroken. People living in this state of hunger are in a constant state of gray—not quite alive, not quite dead. The most maddening thing about this is that these are *farmers* who can't grow enough to feed themselves. If they were plumbers or electricians or bankers, we might understand. However, because these are hundreds of millions of *farmers* who can't feed themselves, it underscores the problem even more.

At the height of the 2009 economic crisis, thousands of people descended on Rome to attend two summits aimed at ending hunger for over a billion people. Four days of intense discussion led to two vastly different outcomes. The primary difference was in who makes decisions about food: **the farmer**, which leads to *food sovereignty*, or **a host of international agribusinesses and food aid providers**, which leads to *food security* through *food aid*.[3]

Food aid and food security have long been used by the development community to ensure that people have enough food to fill their stomachs. Food security focuses on the mass production of grains, oils and sugars by large agribusiness for export and food aid distribution. This often interferes with local markets and further increases food insecurity. Everything that is promised with food security is based on the premise that the Western model of food production, processing and consumption is the only way to feed the world.[4] Like the data produced about genetically modified food safety, the researchers and distributors of information regarding food security have been infiltrated by businesses that profit from the expansion of industrial agriculture. These businesses throw around the word "food security" without understanding hunger or its long-term implications. When hunger begins at an early age, it inhibits development. Within the first two years of their life, malnourished children can have their immune system and brain capacity irreversibly decreased. This stunts a child's capacity to learn, be healthy and provide for themselves in the future.[5] It also further exacerbates the hunger-poverty cycle. In addition to calories, our future food production must include the vital micronutrients people need to be healthy and productive, not to just survive.

When farmers make their own decisions about what, when and how food is grown, eaten and sold, we see vastly different outcomes. Small farmers make up of 52 percent of Africa's work force.[6] Historically, they have grown 150 different crops, 115 of which are indigenous to Africa.[7] By continuing to persuade impoverished farmers to grow one meager yield of a common monocrop, we are limiting their productivity potential and failing to meet their many needs. Putting food-growing decisions back in the hands of the farmers gives them *food sovereignty*,[8] channeling hundreds of years of knowledge about indigenous food and food-growing methods back into their fields. This is fundamental to the development of local, sustainable agriculture systems where both common and indigenous varieties of trees and crops can be grown to feed a growing population. When the farmers have sovereignty, they choose to grow a nutritious set of foods that benefit their families and their farms. They use the diversity of local species to optimize the growing potential of their land. When researchers, input providers and big agribusiness have control, farmers end up degrading their precious lands to optimize the food and financial needs of other people.

During the hot, dry season—a month after I ate the rat—the locusts came. They ate everything that wasn't underground. The community was devastated. But Gueye had planted some beets among the peanuts. For weeks, we ate only beets until our pee turned red. Although beets were one of the few foods I hated, nothing ever tasted so good. After months of eating shades of white, brown and grey, we finally had color in our food. We train farmers in our projects to always have red, orange and green growing on their farms, giving them the micronutrients they need. The agriculture development industry invokes yields and calories to make their case for industrialized agriculture, stating that with population growth, agriculture yields must double in the next three decades.[9] However, focusing on yields and calories does not solve the problem of nutrient deficiencies. We must talk about the micronutrient value in those yields. If you have nothing but corn in the field, then you most likely have only corn on your plate. Vitamins don't grow themselves; we must grow them. In addition to calories, our future food production must include the vital micronutrients people need to be healthy and productive, not to just survive.

**7 Billion People in the World,
4 Billion Malnourished[10]**

1 billion people do not have enough food

2 billion do not have enough of the right food

1 billion have too much of the wrong food

Across the Atlantic, in the inner cities of the East Coast, access to food is plentiful. Convenience stores exist on every corner. When low-income families are hungry, they can afford an overcooked hotdog, a loaf of white bread and a liter of soda. These foods are loaded with calories, but not nutrients. A single mom living and working in the inner city of Washington, DC, is already stretched to the max in term of time and finances. To travel to a supermarket for fresh fruits and vegetables requires two hours and four bus changes with bags of groceries and kids in tow. When she arrives, she finds that her income permits her to buy fruit gummy snacks and vegetable chips but not fresh fruits and vegetables. However, inner cities are not an anomaly; people throughout the U.S. live in what the USDA refers to as *food deserts*. Imagine: the largest producer of food in the world with neighborhoods and towns where healthy food is so scarce that it is labeled a desert. Ironically, some of these food deserts exist in places where mass quantities of food are grown or distributed. The world's largest wholesale fresh produce market is in the South Bronx: 113 acres of colorful, mouthwatering fresh vegetables and fruits, to which local residents have no access.[11] In a nation with food surpluses, it is clear that national food security does not mean individual food security, and it's a long way from food sovereignty.

On the West Coast, California's Central Valley produces over a third of the fruits and vegetables grown in the United States.[12] However, low-income residents in the valley find it far easier to buy peach soda than a peach. Most of the fresh food is quickly and efficiently trucked off, sold in larger cities and sent to processing plants. It returns to them as packaged, processed and fast food.[13] These processed foods not only line the shelves of grocery stores, they are also served up in public school cafeterias across the country, and affluence is not a contributing factor. Not far from where I live, Fairfax County, Virginia, has the second highest median income in the country.[14] Fairfax county public schools are hailed as some of the best; however, the school lunch food choice offerings include delectable items such as corn dog biteables, chicken nuggets, pancakes in a bag and Pop Tarts. Although children have the choice to bring their lunch, many lower income families rely on reduced and free school lunches for their children's nutrition. Thus, the kids who cannot afford nutritious food at home are getting the same sugar-, salt-, preservative-loaded diet at school. This diet leads to obesity, diabetes and other health problems, which in turn leads back to poverty and the continuation of the cycle from one generation to the next. They find themselves stuck in a food system that has been designed to make junk food affordable and available and healthy food overpriced and out of reach.[15]

However, it is not only low income families that are malnourished. Our intensely busy, overworked society has turned to packaged and frozen meals that have the same nutritional value as the plain rice and millet eaten in Africa. The only difference is that our processed foods are loaded with taste: fat, sugar, salt and artificial flavorings. In the U.S., we are only three generations removed from a time when nearly everyone experienced hunger at some point. Now we have millions of people who are obese but malnourished.

These cyclical food and health problems in both urban and rural areas throughout the world have opened up a space for change. Instead of changing the capitalistic conditions of the industrialized food system, we can transcend the system by building a movement within its cracks.[16] In food deserts and poverty-stricken rural areas, cracks in the industrialized food system are plentiful. It is in the cracks that small yet superior food growing systems such as forest gardens can take root.

KILLING MANY PROBLEMS WITH ONE GARDEN

"Danga dof." *You're crazy* is a phrase in the Wolof language that Mariama Ndao heard countless times over two years as she pulled water from an eighty-foot well and carried ten-gallon tubs on her head, over the length of a football field, to water her eight hundred newly planted jujube trees. As Mariama, who was in her mid-50s, carried dozens of tubs of water a day, "You're crazy" seemed to the villagers to be a gross understatement. "They are just thorny bushes you can find in the countryside," they insisted. "Why are you tiring yourself in the hot sun to plant them in your field?" The villagers all knew that the thorny, fruit producing jujube bushes were no longer in abundance. Once covered in trees, Senegal had become a victim of slash-and-burn farming and the cutting down of trees for fences, homes and fuel wood. Land degradation, drought, herds of grazing animals and the need for firewood had left very few natural jujube trees standing. Wangari Mathaai, the tree planter who won the Nobel Prize, said, "Poor people will cut the last tree to cook the last meal."[17] This is exactly what is happening in Senegal, and Mariama was determined to reverse the trend. But in Kaffrine, Senegal, where Mariama lives, the subsistence way of life does not offer much room for vision. If you struggle to eat today, it is difficult to see ahead to next week and nearly impossible to see two years ahead. For decades, Mariama had been stuck in this cycle—having little nutritious food, poor health, no grass for her animals and no wood for cooking. She did what her father had done: till the land, plant peanuts, harvest. Till the land, plant peanuts, harvest. With each year, the land yielded less and less. The two-hundred-dollar pay day for her peanut crop wasn't enough to support her family throughout the year—you can't feed a family for a year with peanuts. For years, she'd been pressured to buy fertilizers and higher yield varieties of peanuts. She'd tried them, but they didn't give her the gains she needed. She couldn't afford vegetables or fruits. Not one of her children had been able to attend school. At night she lay awake thinking, "What will happen to them?" What would her family eat when the dying land's hunger bar ran out? The fear of her children continuing the cycle was greater than the sharp tongue of her friends, so she continued to carry water and care for her trees.

Mariama's friends called her crazy, but her need to support her five children and four grandchildren on a barren peanut field, with virtually no remaining natural resources for miles, had left a crack in the industrialized agriculture system. In this crack she had planted hundreds of thorny jujube trees in rows around her two-acre degraded farmland. As her trees grew, she planted vegetables in their protection. Mariama's vision involved growing what she wanted to grow, and what she needed, on her own land. By planting and growing jujubes, she is expressing food sovereignty. Jujubes are a fruit with a large local demand that always sell out on market days. No one in the United States is growing and exporting jujubes, proving Mariama wasn't so crazy after all. She also knew the lines of thorny trees would someday become a living fence, providing protection from grazing animals, shelter from the hot dry winds, a sustainable supply of firewood and hundreds of pounds of fruit to eat and sell.

Each year in July, as the harmattan winds blow in from the Sahara and the heat rises to bake what the winds do not touch, people wait for the rains. In 2011 and 2012 the rainy season came late and ended early, leaving an estimated twenty million rural residents in the Sahel fearing severe hunger.[18] Food aid organizations scrambled to get ahead of the crisis, providing emergency food security in the form of rice, wheat and millet. But food security alone was not what Mariam sought. It was food sovereignty that she was after. It was the desire to have control over what she grew that gave her the willpower to work hard and ignore the "Danga dof" comments. When her neighbors were worried about food security, Mariama was not worried about how she would support her family. Just three years after planting her first trees, a forest garden had begun to emerge within the protection of her thorny living fence.

The following year, Mariama's vision began to pay off. That June she harvested nearly ninety pounds of pigeon peas. These nitrogen-fixing bushes strengthen the soil while generating both food and income. She sold half for nearly $50 and kept the other half to eat. Although $50 will barely pay my water bill in the U.S., it was enough to lift Mariama and her family above the one-dollar-a-day poverty line for a couple months until the young jujube and cashew trees began fruiting later in the year, bringing her another payday. Her forest garden gives her year-round, individual control of her family's nutrition, income and health. Now the members of her women's group, who once called her "crazy," stand beside her replicating a tree planting strategy that meets all their basic needs, discovering that trees are inseparable from their survival.

GROWING SOVEREIGNTY

Two years ago my eyelids started twitching, and I began having painful back spasms. My doctor suspected it was diet related and a nutrition app confirmed I was magnesium deficient. As I used the app to record my meals and calculate my nutrient intake,

I discovered it was very hard to meet my magnesium needs of around 500 mg per day. Each day I needed to eat fifteen stalks of broccoli, fifteen bananas, or twenty cups of spinach.[19] Even with combinations of these foods, I couldn't come up with these numbers. This made me wonder: if I couldn't meet my micronutrient needs with an entire supermarket down the street, what is Mariama supposed to do in Senegal? Through trial and error I discovered a surprisingly simple three-step process. Find out what you need, grow what you need, eat what you grow. I figured out that I needed magnesium and that there are high levels of magnesium in green leaves (particularly spices) and nuts.[20] I dedicated myself to growing basil, oregano, mint and parsley in my garden. I learned to dry and store herbs and make and freeze sauces like pesto and chimichurris. The nutritional answer for our African farmers is the same: grow what you need, eat what you grow. Then I realized that without a single nutrition class or an app, many of the fruits trees and crops that our farmers in Senegal had chosen to grow in their forest gardens were not only naturally healthy, but contained the vitamins that would meet their micronutrient needs. Across the continent in Machakos, Kenya, our farmers are growing twelve fruits trees that provide micronutrients year-round, one fruit for each month. This underscores the importance of farmers having control over what they grow, eat and sell. When given a choice, they know what is best for them. They don't want to eat what we eat. We don't want to eat what we eat. We just don't know how to eat differently. When I think about the generational knowledge of food in the United States, much of it has been lost. Many of us grew up during a food conversion era when packaged food with additives was promoted as modern meals. There are many cookbooks from the 1970s and 1980s that require canned cream of mushroom soup or cream of celery soup or Velveeta cheese for nearly every recipe. For three to four decades, most of our meals consisted of meat with a side of canned or frozen vegetables, salads drenched with creamy sauces and buttered processed bread. We have at least two generations of "lost food" people that must relearn how to cook and how to eat.

U.S. farmers were not given a choice. In the eighties when everyone was losing their farms, the government wasn't handing out subsidies for growing spinach or basil or broccoli. It was all about corn. Now three decades later we are not eating and thriving with all the micronutrients and vitamins we need. We are primarily made of corn.[21] The same thing is poised to happen in Africa with AGRA's new Green Revolution. Farmers are not being given a choice whether they want to grow mangoes and rattan and tomatoes and cashews. No one is handing out subsidies or seedlings for these crops. Instead farmers have pressures from all angles telling them they can get free seeds, reduced-cost fertilizers, government subsidies and nonprofit funding if they grow monocrops.[22]

When choices are limited, people's potential is limited. When a farmer plants only one large crop and harvests one large crop, he or she has two choices: store it or sell it. Storing food is not easy and poses many risks. I know, because I wrote long technical manuals on post-harvest handling of cereal crops in East Africa. If I were a farmer, the last thing I'd want is to be responsible for storing my own food for weeks or months. Rats, mildew, temperature and humidity fluctuations, insects and lack of proper materials and infrastructure make it extremely challenging to store food for long periods. Few farmers have access to centralized storage and grain warehouses, so most must store it in their household. Even if it is stored, humans cannot live on one grain alone. Some of it must be sold for income to purchase other sources of nutrition. If farmers can't store their grain crop, then they must sell it quickly. Unfortunately, there will be thousands of other farmers selling their crop at the same time, forcing them to sell at the lowest price of the year. They are often desperate for food and cash, further diminishing their bargaining power. This means that farmers are damned if they do (sell) and damned if they don't (store). When looking at it from this perspective, it does not make sense to encourage farmers to replicate or expand a monocrop agriculture system in which they will always lose. A better option is to grow and sell more profitable and nutritious foods, while continuing to keep grains as a smaller part of their growing system.

Although we can't fight big agriculture, we can take advantage of all the cracks in industrial agriculture's system and begin to convert portions of fields to forest gardens. We are not insisting that developing a forest garden is a simple process. But with time it becomes less labor intensive and more rewarding than traditional farming, because each year farmers build on the previous year instead of starting over with a barren field. Developing a forest garden is not an easy task, and it requires a great deal of hard work and commitment, particularly in the first four years, as the trees are becoming fully established. We like to think of the forest garden training process like getting a college degree. For four years, a student works two jobs and studies, stays up late, eats very little and lives very cheaply. But after that sacrifice, the student has something that will benefit him or her forever, something no one can take away. This is the same for the farmer who plants a forest garden. During the first couple years, there is a lot of work to plant and care for the trees, but then the garden begins to slowly produce. After four years, when the forest garden becomes fully established, it provides a lifetime of growing rewards.

Mariama is just one aging woman in Senegal. Her trees alone will not stop climate change, nor will they replenish Senegal's falling water table or stop desertification. But they have created a sustainable forest garden to nourish her family and they have begun to end the cycle of poverty and hunger.

Mariama has planted a seed of hope in a place where hunger is rooted deep. With that hope other small farmers are beginning to plant trees. Combined, these forest gardens CAN change the course of environmental degradation. In abundance, thousands of farmers' forest garden trees CAN restore groundwater tables. In abundance, they CAN provide food sovereignty for an entire region. In the workshops we conduct in Senegal and across Africa, farmers like Mariama are learning to plan out the year, identify lean times and ensure something nutritious can be harvested every month of the year, meeting their families' needs first. Day after day, like the imperceptible growth of a tree, a movement is forming; Mariama is no longer just one crazy woman with a couple thousand trees on a barren lot. She is the root of a vision beyond the cycle of poverty and hunger to a place of sustainable independence. To a place of sovereignty.

CHAPTER 8

WEATHER OR NOT

CLIMATE CHANGE

We lost in seventy-two hours what we have taken
more than fifty years to build.

—HONDURAN PRESIDENT
CARLOS FLORES
Following 2005 Hurricane Mitch

BRINGING STALE BREAD TO THE POTLUCK

On May 10, 2013, I got into my car, dropped my kids off at school and began my commute to work. I was listening to my favorite morning radio show. As I stopped at the traffic light the comical banter was temporarily interrupted by an overview of the day's news headlines,

"Death Toll from Bangladesh Building Collapse Tops One-Thousand. The Man Accused of Holding Three Women Captive in Ohio Could Face the Death Penalty. Republicans Block

Confirmation of Obama's Pick to Head the EPA. Minnesota House Passes Bill to Legalize Same-Sex Marriage."[1]

'For the First Time, Earth's Single-day CO2 tops 400 ppm'[2]

Staring ahead with horns blaring behind me, I missed my green light. *Four hundred parts per million (ppm)!* I remembered Al Gore's warning of 350 ppm as a safe target, but we had surpassed that in the late 1980s. For the science community, reaching 400 ppm was a blow. But there was no indication on the radio or from the blaring horns behind me that anyone understood the significance of this news. I took a deep breath and tried to ignore the obscenities spewing out the driver's window behind me; I knew he was not cursing the 400 ppm. When the light changed I kept driving, expecting that at any moment the radio's Emergency Response System would kick in saying, "This is not a test . . ." I had imagined that if the atmosphere ever reached four hundred ppm that cities would begin holding emergency preparedness meetings (like the kind I've seen in Jamaica), coastal communities would begin to design flood walls and local governments might order restricted energy use (like California has done with water during its long drought). I thought we all might park our cars along the interstate and walk away from them or run to the store to stock up on supplies, planning for the worst, like we do for ice and snow storms. But, of course, none of this happened. It was just another headline, lasting less than a second, but its impact will affect us for generations.

Prior to the industrial revolution, we were at 280 ppm and the global temperature was nearly 1 degree Celsius cooler.[3] However, since then we have been burning fossil fuels and putting more CO_2 in the air than plants can take out. In addition, the destruction of forests and conversion of forests to farm land has increased the amount of CO_2 going into the atmosphere. When trees are burned, cut down and removed they release CO_2. Likewise, this reduces the amount of land area (forests) that can absorb carbon. Each year in September, scientists find that CO_2 levels dip to their lowest point, due to a long summer of forests absorbing carbon in the Northern Hemisphere.[4] This dip has been observed every year since its spike to 400 ppm in 2013.[5] However, in September 2016 the low point was above 400 ppm, which indicates that this is the new norm, and it will only grow higher.[6] The last time the Earth was at 400 ppm, humans did not exist. The last time the Earth was at 400 ppm for a sustained period of time was fifteen to twenty million years ago, a time geologists call the Early Miocene.[7,8] Sea levels were twenty-five to forty feet higher and the Artic was free of ice year-round, with global temperatures three to six degrees warmer.[9] So why hasn't this happened yet? Well, the Earth is playing catch up. In the past, CO_2 levels rose and fell over millions of years. We have had a 100 ppm jump in less than one hundred years.[10] In Earth time, one hundred years is like a few seconds. The Earth only realized a few seconds ago that CO_2 levels are rising, but already it is responding. Ice is melting. In 2007, the Arctic's Northwest Passage became navigable for the first time.[11] In 2014, a year after the earth's atmosphere reached 400 ppm, two independent studies concluded that the Western Antarctic ice sheet is in irreversible retreat.[12] With the melting of polar ice, seas are rising. In the last few years ten thousand people from the Marshall Islands have migrated to northwest Arkansas to escape rising sea water.[13] However, islanders are not the only ones at risk—one in ten people worldwide live in low elevation coastal areas that will be impacted by rising seas.[14] With the warming of air, the intensity and frequency of floods and droughts are increasing. Think of clouds as giant sponges moving over the land soaking up moisture. Warmer air can hold more water vapor than cooler air, so the land becomes drier and drier before the sponges get full. When they are full, this increased moisture is dropped in the form of rain or snow, many miles from where it was soaked up.[15] With this warmer air, temperatures in parts of the world are soaring—for two weeks at the end of May 2016, temperatures in India reached an all-time high of 131°F. It was so hot that streets liquefied; people trying to cross the street got stuck in melting asphalt.[16] Climate scientists predict that increased heat stress and drought, rising salt water and loss of pollinators will severely inhibit plant growth in equatorial and tropical zones.[17] Increased storm intensities, flooding and drought have led to increased structural damage. Between 2013 and 2016, the U.S. had thirty-nine weather events that each caused over one billion dollars in damage each.[18]

In the summer of 2016, the hottest year on record, I traveled a great deal speaking about our work and meeting with people and donors. There is always a big misunderstanding between climate (which is a long-term global change) and weather (which is a short-term local occurrence). However, in every town and city I stopped in people were eager to talk to me about changes they are experiencing locally. The summer started early with an unusually hot and dry May in Canada, where temperatures reached 91°F, sparking Alberta's largest wild fire.[19] As it burned for two months the fire grew to the size of Delaware and smoke affected people from Michigan to Florida. It was even detected across the Atlantic in Spain. Farther south, May brought Maryland seventeen days of rain that dumbfounded local weatherpersons. While California was in the midst of its fifth year of drought, May dumped twenty-two inches of rain in the suburbs of Houston in less than twenty-four hours.[20] As June approached, I was sent a message from my favorite orange producer in Florida. They canceled this year's orange crop due to the rapidly spreading citrus greening disease called *huanglongbing*. I learned that this bacterial infection, which is spreading quickly in Florida's warmer drier weather, has no cure. It has infected 70 percent of Florida's orange groves, making the future of orange juice look grim.[21] This shows once again what happens when things are grown in monocultures, even trees. On July 6, news headlines in Miami read, "Sahara Dust Storms Reach Gulf Coast." I followed this dust back through the hurricane factory to visit our projects in West Africa. There we can no longer predict the growing season—floods come at unusual times and drought plagues the typical rainy season. Across the continent, our team working in East Africa now plans for drought every year. Back in the United States in August, I spoke with a woman in western Massachusetts, who said that the rain had not crossed the Connecticut River all summer. In an average year, eleven inches of rain fall in the summer months, but this year the clouds would come close and then just disperse.[22] While the Connecticut River flow was waning in August, Baton Rouge, Louisiana, got over twenty-five inches of rain in just eight days, the most rainfall on record, giving them a total of over ninety inches for the year. Later as we rounded out 2016, the state was pummeled by thirty-eight tornadoes in less than forty-eight hours.[23] These weather events may be local, but when connected we see a pattern of changing extremes, anomalies in every state and every country that have a local and global impact—anomalies that affect people's lives, crops and all life that is trying to adapt.

In 2006, Al Gore's movie *An Inconvenient Truth* cast light on the climate change dilemma. At the end of the movie there was a list of things we could do to curb carbon emissions and prevent the advancement of climate change. At the time, just ten years ago, people were encouraged to unplug their appliances, install energy efficient lightbulbs and make sure the tires were properly inflated.[24] All simple things that could be done to reduce energy use and improve fuel efficiency. Now, we are way beyond that.[25] Current trends and future forecasts indicate that we are close to a path of no return. Significant action must happen now.

In an effort to stir up action, 195 nations signed the Paris Agreement at the Paris Climate talks (COP 21) in 2015. They all committed to a vague nonbinding agreement that set lofty goals, including keeping the worldwide temperature increase to well below 2 degrees Celsius.[26] To achieve this temperature goal, all parties are left to determine their own goals with the aim to reach global peaking of greenhouse gas emissions "as soon as possible." Or, as John Cassidy so bluntly put it in a *New Yorker* article, "It's a bit like a potluck dinner, where guests bring what they can."[27] As the world's leader, we would expect the United States to bring several large gourmet dishes to the potluck. However, the United States federal government's interest has bordered somewhere between reluctant and negligent in its response to climate change. In fact, for years we've been showing up to the potluck with a few pieces of stale bread, hoping that developing countries will accomplish what we can't—a switch to renewable energies.[28]

As the earth reaches its maximum for CO_2 mitigation, countries are desperately searching for areas in which to reduce greenhouse gas (GHG) emissions. The farming sector is a big one. As one of the largest emitters of GHG, the farming sector not only provides great opportunities for decreasing CO_2 emissions but also gives us an opportunity to capture the CO_2 already causing climate change. Land use change including deforestation to make room for crop production accounts for 15 to 18 percent of total GHG. Plowing, fertilizing and harvesting is responsible for another 11 to 15 percent. Food distribution including transportation, packaging, processing, retail and waste is one of the largest emitters, contributing 17 to 24 percent of total GHG emissions.[29] This means our current food production and distribution system accounts for about half (43 to 58 percent) of total GHG. We often focus on transportation as an area where we can significantly reduce GHG emissions, but in reality, individual transportation only accounts for 6 percent of CO_2 emissions.[30] The real impact lies in buying local, sustainably produced food; or as the Food and Agriculture Organization (FAO) calls it, food grown through Climate Smart Agriculture (CSA). CSA aims to sustainably increase agricultural productivity, biodiversity and incomes; adapt and build farmers' resilience to climate change, and reduce greenhouse gas emissions from farming systems.[31] Although CSA was initially intended to define and develop organic, sustainable agriculture approaches, the term "Climate Smart Agriculture" has been hijacked by others with entirely different intentions. McDonalds and Walmart have announced their commitment to Climate Smart Agriculture, while Monsanto and fertilizer giant Yara use the term to promote their climate-resilient genetically engineered seeds and synthetic fertilizers.[32,33] Industrial agriculture has a two-pronged Climate Smart Agriculture approach: 1) increase production in the United States for export to countries that cannot grow enough and 2) create a green revolution of agriculture intensification in Africa. Both prongs are fueled by the sales of their products: GM climate smart seeds, herbicides and fertilizer. Climate change models do show that that there could be an extended growing period in the Midwest due to climate change. However, increased heat and drought stress on crops and the health of the soils could be limiting factors. A longer growing season is also likely to increase pests and weeds.[34] In Africa, climate models reveal that the continent may have a 15 to 30 percent reduction in agriculture productivity due to increased temperatures, intense storms and unpredictable rainfall patterns.[35] Farmers must transform their fields into biodiverse, flexible agriculture systems to adapt to these changes. Relying on industrial adaptation strategies by increasing use of GM seeds, fertilizer, herbicides and irrigation will only increase emissions and the environmental degradation of the land, exacerbating climate change effects and inhibiting future crop growth.

Because Climate Smart Agriculture is being used to define two very different systems, we must look at how CSA crops are grown, how far they were transported and the end use of the crops. Grain produced in the Midwest is not only for human and animal consumption. Fully 38 percent of the

U.S. corn crop is used to produce ethanol.[36] The EPA mandates that each state meet ethanol/gasoline blend quotas. To achieve this, most states sell E10 gasoline, which means 10 percent of the fuel is ethanol (94 percent of this ethanol is derived from corn).[36,37] Although this aims to reduce the CO_2 emissions from petroleum-based fuel, the amount of CO_2 that is emitted in the process of growing, irrigating, harvesting and processing corn for ethanol is not negligible. In fact, the CO_2 that is emitted from producing ethanol is almost equal to the CO_2 savings from its 10 percent contribution to the fuel source.[38] This kind of illogical agriculture loop leads to madness among those who understand it and confusion for those who don't. Why would the 10 percent ethanol mandate exist if it wasn't reducing CO_2? If we look at the other types of renewable energy that may be produced on the same land, we can find some surprising energy and financial yields. This chart represents the BTU hours of energy that can be generated on one thousand acres of land in one year, the retail value of that energy and the energy return on energy invested (meaning energy out versus energy in).[38,39]

ENERGY SOURCE	RETAIL VALUE	ENERGY YIELD	ENERGY OUT VS. ENERGY IN
Corn Ethanol	$612,000	26 billion BTUs	1.5:1
Wind Power	$8,000,000	224 billion BTUs	50:1
Solar Power	$41,640,000	1.18 trillion BTUs	25:1

Looking at the numbers it appears that the only beneficiaries are the businesses providing GM corn seeds, herbicides and fertilizers. We think, *there must be a logical explanation*. In the absence of one, we can conclude that the reason for an ethanol mandate is not to reduce CO_2 emissions, but solely to provide another market for corn.

UNOBTAINIUM

In 1928, the Kellogg-Briand Pact was signed. This "Treaty for Renunciation of War as an Instrument of National Policy," as it was officially titled, was signed by fifteen countries including the United States, France, Germany, Italy and the UK. It sought to outlaw warfare as a means of settling territorial disputes. Although the treaty lacked an enforcement mechanism, its signature was hailed as a historic turning point in world peace. A year after its signing, U.S. Secretary of State Frank B. Kellogg, who helped design the treaty, received the Nobel Peace Prize; within five years, sixty-five countries had signed the agreement. Within a decade, the world was at war.[40]

Parallels between this historic treaty and the Paris Agreement Climate Initiative cannot be ignored. Without binding commitments for CO_2 reduction, action will be slow. As humans, we like deadlines and challenges. The more difficult and unobtainable, the more we want it. Fossil fuels were once an unobtainium, a resource that seemed impossible to extract and process. Now their extraction and use has become as common as pumping and purifying water. We've drilled, fracked and removed mountaintops, anything to obtain the unobtainium. But we haven't yet tackled renewable energy on a massive scale, perhaps because we've never really been challenged. I know we can do it. Working closely with businesses who help us plant trees and develop forests gardens, I know that it is often business that finds a way through the fog. In fact, it is individual cities and businesses that are taking the most progressive and tangible actions toward developing renewable energy. I DARE YOU to make renewable energy your new unobtainium.

For those who don't believe in climate change, who think that climate science is hocus pocus, I will not use this section to convince you otherwise. Although I am an absolute believer in climate change, I do not believe in the fraudulent treaties that have been hailed as ground breaking climate change agreements. Rather than focusing on climate change itself, a problem of planetary proportions that can make individual efforts seem futile, I ask that you consider creating lofty inspiring goals that can shift the focus from a looming climate disaster to building an infrastructure legacy, moving us from the end of oil to the beginning of a new economic era for the United States.

This may seem like an impossible feat, but only 150 years ago, replacing whale oil felt like an impossible feat. Whale oil was one of the first globally traded energy sources. Throughout the 1700s and early 1800s whale oil was the fifth most profitable industry in the United States. At the peak of whaling, 735 out of 900 whaling ships world-wide were U.S. ships.[41] In 1859, the first oil well was drilled in Pennsylvania and kerosene from oil replaced whale oil.[42] However barbaric the slaughter of whales for oil may seem, it took 150 years to replace whale oil with ground oil. Now 150 years have passed since we began using ground oil; isn't it time to replace ground oil with something else? This question is particularly pertinent as our extraction of ground oil becomes more barbaric. When I first heard about removing oil from the tar sands of Canada, I imagined a desolate, desert-like area with pools of oil just below the surface of the sand that could be extracted and pumped. What's the harm? We already pump oil from all over the world. However, I discovered that the tar sands are not a desert like area at all, but in fact are located under the largest ecologically intact boreal forests in the world, comprising 54 percent of the total boreal forest on the planet.[43] Prior to accessing the oil, the land is stripped of these trees, then dredged and pressed until it looks like a landscape from *Lord of the Rings*.[44,45] The slaughter of the boreal forest and its species to remove oil from the tar sands is no less humane than the extraction of oil from whales.

Eighty years ago, our nation was in shambles over the Dust Bowl and the Great Depression. At that time, President Roosevelt saw a way forward through infrastructure and art, building and documenting this historic moment to inspire Americans for generations. Fifty years ago, our nation was ablaze with anti-Vietnam-War protests, the looming Cold War and the Civil Rights Movement. President John F. Kennedy saw a way forward by going to the moon. Kennedy realized that shifting the focus from an internal struggle to the lofty and inspiring goal of sending the first man to the moon meant shifting the American mindset from war to science, changing the focus from revenge to education.[46] Today America is divided over energy, job creation and which direction to take the country. In the midst of this divisive climate, President Trump has an opportunity to see a way forward with something bigger than going to the moon, a new energy future and agriculture revolution.

The world's ability to see the first moon landing via television launched America to the forefront of progress. It made us the experts. Today European nations are becoming renewable energy experts. Denmark's electrical grid is run on 100 percent renewable energy.[47] Sweden has declared that it will become the first fossil fuel-free state.[48] Does the United States really want to be surpassed by European nations in this technological advancement? It is time for Americans to become the experts again, experts at reinventing energy systems and sustainable agriculture production.

The only massively exported technological change that has happened in the last fifty years has been in communications technologies. The advent of the Internet, cellular communication and smart phones spread so rapidly, that now, 65 percent of the global population own a cell phone.[49] More people have cell phones than have electricity and running water.[50] Why? Because this technology is innovative, portable, relatively inexpensive and works everywhere. Imagine if we could create a renewable energy system that didn't require billion dollar grids, that could be used anywhere, that could be reproduced and implemented and spread throughout the globe within twenty years. Imagine if we could transform agriculture around the world and in the United States so that forest gardens became as prolific as convenience stores, with one on every corner. Imagine if millions of farmers were growing fresh produce in their communities, making agriculture once again a part of our culture. That is a legacy. The effect on America's wealth, well-being, and position in the world would be enormous. The well-being of people worldwide would improve. And as a sidebar, the impact on the environment and the climate would be positive.

As we face the massive endeavor of creating employment, rebuilding American infrastructure and building sustainable energy systems, I ask you to take this on, not for the sake of climate change, but for the sake of legacy. Just like you, our farmers are not planting trees for the sake of climate change. They are planting trees for every other reason in this book. They are planting to have food year-round and energy and fodder for their cattle. They are planting trees to keep their land from blowing away, to keep their water from disappearing and to have a steady income. Although they are experiencing increased flooding, droughts and storms, they do not lie awake at night and think of climate change. They lie awake at night and think the same thing you think: "How can I use my resources to make more money?" Although they are only looking for a few dollars more a day and you may be looking for a few hundred dollars more a day, the desire is the same.

GROWING DILLERS,[51] DOLLARS, BREATHERS AND SOLDIERS

Maybe the best way for us understand the threat of the coming decades is through apocalyptic Hollywood movies. In these movies, human extinction is often brought on by alien invaders, super germs, meteors or massive storms. However, in my son's favorite movie, *Pacific RIM*, earth's invaders, the *Kaiju*, did not come from the sky but from the depths of the Pacific Ocean's Mariana Trench.[52] As scientists learn more about the global effects of climate change, we can look to increasing CO_2 in the atmosphere as the cause, but it is to the oceans we must look for the apocalyptic changes that are coming, including rising seas, dying coral reefs, the loss of millions of species and enormous storms. While oceans can cause massive havoc, they are also our largest carbon sink, absorbing 25 percent of all CO_2. However, when CO_2 dissolves in salt water it causes the ocean to become more acidic. In the last two hundred years the ocean has become 30 percent more acidic, dissolving the shells of some of the most important carbon absorbing organisms.[53] In addition, as ocean waters warm they stratify, decreasing the ability for layers to circulate and absorb carbon.[54] With CO_2 emissions on the rise we need more organisms that can absorb carbon. Trees and organic, rich soils are the only answer.[55]

But how many trees does it take to change a lightbulb? Or a climate? In 2013, in response to a group wanting to know the answer to this question, Thomas Crowther from the Yale School of Environmental Science started by determining how many trees there were on the planet. He had assumed that this was something already known, but in fact it wasn't. Using computer models, inputting data from four hundred thousand forest plots where trees were hand counted and combining this with satellite imagery, Crowther and his team determined that there are about 3.04 trillion trees on earth.[56] Comparing this with other historical data, they estimated that since the dawn of civilization, about half of all the earth's trees have disappeared.[57] However, if you look country by country, far more trees have disappeared in some areas than others. Once covered with trees, as of the late nineties, seventy-six countries had lost all of their frontier forests (original tree cover). This includes most European and

East African countries and all of North Africa and the Middle East.[58] Some of these countries have recovered with conservation policies and massive tree-planting efforts but others have not, and still others continue with rapid deforestation. Trees once provided natural protection, acting as dug-in soldiers shielding countries from typhoons, hurricanes and monsoons. They covered the countrysides, cooled the land, brought the rain and channeled excess water back into the ground. They stabilized the soils, holding them in place against wind and rain, preventing sandstorms and mudslides.

As soldiers and breathers, trees provide both CO_2 reduction and mitigation, serving as a nonpartisan weapon that is exempt from climate politics, whose beneficial existence is not subject to scientific evidence or debate. So their value should be recognized, right? The United Nations aims to put a value on trees as breathers through their Reducing Emissions from Deforestation and Degradation (REDD) program, which involves a payment for carbon storage, leaving trees untouched.[59] Although this provides a small sum of money, like peanut farming it does not provide everything else people need. Farmers need wood, money, food, fodder and purpose. A farmer cannot plant his land to trees and just let them breathe in CO_2. He needs that land and what it can produce. However, if trees can be incorporated into that model, then we are growing wood (dillers), as well as money and food and soldiers and breathers on one plot of land. This combats climate change while providing all elements needed to lift people out of poverty.

In October 1998, just before the widely celebrated "Day of the Dead," Hurricane Mitch hit Honduras. It was the biggest storm the country had seen in a century and their fallen tree soldiers were unable to defend them. The large-scale deforestation of the 1980s and 1990s had left the mountains bare of trees. When the rain poured down, there was nothing to hold the soil in place and the mountains slid into the valleys of cities. Twenty percent of the country became homeless, 70 percent of the crops were destroyed and more than eleven thousand people were killed.[60] Four billion dollars of damage was done. Devastated, Honduran President Flores declared, "We lost in seventy-two hours what we have taken more than fifty years to build." [61] In poor countries throughout the world, there is an expectation that, in time of disaster, the international community will generously "step up" and save them. But underneath that expectation lies an intuition, a knowledge that climate conferences, food programs and disaster aid will not save them from future storms, chronic food shortages, rising seas and waning fuel supplies. Farmers in developing countries know that at the end of the day they are still poor and tied to the land. By transforming their land into forest gardens, they develop their own institutions—institutions without walls or fees. Institutions that can support their families and provide resilience to the extreme weather and food shortages that lie ahead.

CHAPTER 9

GREENER GRASS

EMIGRATION

The dreams of youth know no borders.
—UNKNOWN

I HAVE A DREAM

It wasn't the day Omar caught the rat, but it was one like it. In the peanut field, where the ground was as hot as the sun, there was nowhere to escape but inward. As Omar and I rested in the shade of the last dying bush mango tree, we shared our dreams for the future. Omar and I were born worlds apart but we were both young men with no responsibilities but ourselves, single, free and in search of adventure.

"What are your dreams?" he asked me.

"I'm living it," I told him, spreading my arms out to showcase my calloused, dust crusted hands and feet.

He laughed, "You are a crazy *toubab* [white man]. Your dream is what we are all running from."

"You have a good life here," I told him. "You have people that love you, peace, laughter and happiness. You just need to make a little more money."

"I know a way to make more money and then bring it back. There is a boat that leaves Dakar going toward Spain," he said very matter-of-factly. "Before it reaches Spain it is intercepted by the Red Cross. When they take you to the hospital on land, you jump out of the window and run. Then you are in Spain. Free to make money and send it home."

This time it was my turn to laugh, but I didn't, because he was so serious and I knew that for him, it wasn't a joke. In the Wolof language, there is even a cool word for people who earn money abroad and send it home—*modo modo*. The only two concrete houses (not mud huts) in Omar's village were built by *modo modos*, two guys who spent years in the kitchen of an Italian restaurant in Lexington, Kentucky. Now it is much easier for youth to go to Europe than the United States. Omar had tried his hand as a fisherman in Senegal's capital city of Dakar but the over-trawled and overfished shores yielded less than his dying peanut farm.

Omar's story is common for young men in rural Africa. Young men don't wake up one day and leave the village for Europe. They start by going to the city during the dry season to look for work. If they are lucky they can find day labor digging ditches or making bricks. If they are really lucky they may land a night job as a guardian. However, most find very little work because there are thousands of other young men looking for the same thing. What they do find are stories. Stories of how to escape to Europe. Stories of how easy it is. Stories of the riches and the ease of life that can be found by going north. As older men die, men in their midlife leave to find work in the cities and young men leave to go north, women and children are left in the villages to tend failing cropland, vulnerable to theft and violence. Trapped between a desert and a cesspool, families must choose between a failing farm or a failing city—either decision results in misery.

In 2003, a former Peace Corps volunteer and friend of mine was building a health clinic in the village of Bassiknou on the border between Mauritania and Mali, half way between Timbuktu and Nema. While there she met a recently widowed woman, Hadijetu, and her daughter. Hadijetu didn't think she could survive with just a few chickens and a small plot of hardening land with a few dying palm trees. She claimed to have relatives in the capital city of Nouakchott and wanted my friend to buy her house so she would have money to live in the city. Hesitating, my friend warned her of what the city held, but Hadijetu insisted. Hadijetu gave my friend a handwritten deed to her adobe house and my friend gave her $800. They loaded up her meager belongings on the back of a pickup and drove three days to Nouakchott. When they arrived, my friend took Hadijetu and her daughter to *Kebbah*, the place her relatives lived. *Kebbah* is the *Hassaniya* (Arabic dialect) word for garbage and that is what this slum had become. It was built on land that was below sea level near the beach. The government had built a concrete wall to hide its horrors from passersby. For more than a square mile, *Kebbah* was filled with people living in the worst conditions my friend had ever seen, and she'd seen a lot during her seven years in Africa. Scraps of cloth hung over poles as makeshift tents for shelters. There were no latrines, so a mounded border of garbage and human waste had accumulated around *Kebbah*. Between the tents were cesspools of salt water that seeped up from ocean and the only fresh water source was a truck that came by twice a week to fill jerry cans at the cost of one dollar for five gallons. Three months after arriving in the city, Hadijetu showed up at my friend's door in Nouakchott, penniless. She wanted her house back. "Dying in the village is better than living here," she said. My friend gave her back the deed to her house and some money to start again. She had been safeguarding it, knowing Hadijetu would be back.[1] However, many people who find out the hard lesson of migrating to the city have nothing to return to.

For decades, a global forced migration has been occurring as millions of families leave behind their failing farms and move to the hopelessness of urban slums. They don't leave by choice. They leave for all the reasons described in the previous chapters. They leave for fear of tomorrow. Migration is often thought to be a new beginning, but it is actually the end of the road. It is the final result of deforested land, degraded soils, transformed seeds, sprayed poisons, depleted water sources, biodiversity loss and the onset of climate change. With their land, their home and their way of life eroded, there is nothing left to support farming families. Destructive agriculture practices destroy the natural resource base, and that in turn makes the land inhospitable for agriculture. So farmers move, in great numbers, to growing cities with failing infrastructure, no industry and few jobs. Furthermore, with the new risks of climate change, people are becoming more vulnerable in the cities. Slums are often erected in low lying areas, on coastlines and in flood zones. As saltwater invades drinking water supplies, water access will become increasingly difficult in these places. Lack of sanitation, flooding, storm surges and inadequate housing put people living in slums at great risks of death and spread of disease.

In Sub-Saharan African cities, two-thirds of the urban population lives in slums.[2] In these poverty stricken conditions, children and women are at the greatest risk and often become victims of human trafficking. The United Nations Office on Drugs and Crimes indicates that human trafficking is the fastest growing industry in the world and the third most lucrative criminal activity, topped only by the sale of drugs and guns.[3] Eighty percent of these victims are sold into sex slavery and over half are children.[4] Throughout Africa, women always say to me, "Here is my baby, will you take it with you?" Although they are joking, underneath the joke is a thread of hope that I will, not because they do not love their children, but because they always want something better for them. But for the children who actually are taken, what awaits them is seldom something better. Orphaned children face even greater risk of being trafficked. The AIDS epidemic that swept across Africa took the lives of an entire generation in some communities, leaving behind millions of orphans. Worldwide, twenty-five million children have been orphaned due to AIDS and 85 percent of these orphans live in Sub-Saharan Africa. Orphaned children in cities, living with relatives or on the streets, are the most vulnerable to being sold or tricked into trafficking.[5]

As more and more people make this rural-to-urban migration, cities are growing beyond their capacity. The UN predicts that within three decades nearly 70 percent of the world's population will be living in cities.[6] Historically, and even in many Asian nations today, mass rural-to-urban migration has been coupled with industrialization that has led to sustained economic growth. However, Africa has been an anomaly. Although Sub-Saharan Africa has seen some of the highest rates of urbanization in the world, it has urbanized under negligible industrial growth and stagnant or declining economic growth.[7] Between 60 to 80 percent of the population is involved in the informal economy.[8] Nigeria's capital of Lagos is one such city. Lagos has a population of fifteen million people, double the population of New York City, yet two out of every three people live in slum accommodations with no access to clean water, sewer or electricity.[9] Not only is the government unable to develop new infrastructure to accommodate the pace of growth, it is also unable to maintain existing infrastructure. Much of the water and wastewater infrastructure in Lagos has not been operational for the last decade.[10] Half of Nigeria's population of 150 million people live in cities, yet 70 percent of the country's land is arable agriculture land. This land could be used to feed its population, relieve the pressures on the city and grow its economy.[11] By leaving this land dormant and inactive there is a great risk that it will fall prey to government confiscation and leasing to foreign investors. The great tragedy underway in Africa is foreign acquisition of farm ground for growing food for export or resale within countries.[12] This results in former farmers, now city dwellers, paying high prices for food grown on the land they left behind. Over half of all large-scale land acquisition by foreigners is in Africa. "The high levels of interest in acquiring land in Africa appear to be driven by a perception that large tracts of land can be acquired from governments with little or no payment."[13] Much of this land is or will be farmed using Western industrialized agriculture methods. Big agribusiness argues that sustainable farming is too laborious. However, every African nation has the labor to make sustainable farming successful. The International Monetary Fund estimates that by 2035 more children in Africa will be reaching working age than all other countries combined.[14] What will they do? How will they feed themselves? How will they occupy themselves? What conflicts and civil unrest will result if they cannot find employment, fulfillment and hope? Would it not be better for families to sustainably farm their own land rather than work as serfs growing food for export? Would it not be better for families to grow their own nutritious vegetables, fruits, fodder and wood rather than scavenge in city slums and pay high prices for rice grown on their land?

Farmers cannot find employment fulfillment and hope by migrating; in urban areas they only find cities of wrath. Even greater political and economic unrest will ensue if more people are forced off their lands. This scenario silently played out in Syria before the world realized it. What was originally thought to be an Arab Spring movement gone awry was actually a drought-induced civil war that originated fifty years prior with the mismanagement of agriculture land.[15,16] In the sixties, pastoralists along the Euphrates and the Khabur rivers were prohibited from grazing and the land was used for intensive

ONE SHOT

wheat and cotton farming.[17] In addition, the number of wells
nearly doubled in the first decade of the new millennium,
increasing from 130,000 to over 230,000.[18] From 2006 to 2010,
Syria suffered a massive drought that destroyed their intensive
farming and unsustainable irrigation practices.[19] During this
time, 75 percent of farmers experienced total crop failure and
80 percent of livestock died. Around 1.5 million people were
displaced from rural areas into the cities. This loss of livelihood,
internal displacement and failure of the government to respond
is thought to have created the civil unrest that led to the civil
war, and later the mass migration to Europe.[20] Had the farming
systems been comprised of more adaptable and diverse crops and
trees, farming families could have had more resilience during the
drought and the impact may not have been so great. Syria's rapid
descent into chaos is a modern-day warning for all of us. Syria
was not a developing country; it was a beacon of success in the
Middle East. A country filled with well-educated, middle-class
entrepreneurs, skilled tradesmen and professionals. Now from
Syria's farmlands to its city centers, only ruins remain while its
people try to pick up the pieces and start again in foreign lands
where they feel unwanted. Today, hundreds of countries are
on the edge, just a couple droughts away from a civil war and
large-scale migration.

Although the Syrian refugee crisis occurred on a much
larger scale and in a shorter timeframe, migration from rural to
urban areas has been playing out throughout Africa for the last
three decades. This mass rural-to-urban migration has created
a cauldron of civil unrest boiling in numerous African cities.
Explosive population growth coupled with no job prospects and

constant imagery of what the West holds has left many African youth with only one dream—migrating to Europe.

ALL ROADS LEAD TO TRIPOLI

Between 1983 and 1985, more than four hundred thousand northern Ethiopians died during one of the worst famines in recent history.[21] The famine was brought on by a combination of drought, civil war, confiscation of land and government policies,[22] conditions not much different than those in Syria. By the time the world reacted, it was too late. If the millions of Syrians who left their country over the last two years had waited for the world to react, they too might have starved. The difference today is the existence of the Internet and social media. Not because they aid in unilateral decision making or reaction time, but because today, people in need of escape are just two *Viber* or *What's App* calls away from a human smuggler.[23] Between 2014 and 2015 nearly two million Near- and Middle-Eastern people left behind everything, took dingy boats on choppy seas, walked thousands of miles across entire countries, spent days without food, slept in tents in the height of Europe's winter, climbed through razor wire and waited endlessly for hope that came for some and didn't for others. Half of them were from Syria, the other half were from Iraq, Pakistan and Afghanistan.[24] Statistically, migration to Europe has now reached the level seen during the "Great Wave"—the highest period of immigration to the U.S.[25]

The massive scale of this migration made headlines, but under the radar, where the events in Africa often lie, a similar mass migration began with the fall of Libya. In previous decades, Sub-Saharan Africans would try to reach the Canary Islands off the coast of Mauritania or mainland Spain off the Coast of Morocco.[26] However, with the collapse of the Libyan regime and the chaos that ensued, the borders became porous, giving migrants the opportunity to hire smugglers to take them across the desert and Mediterranean to Europe. The city of Agadez, Niger, historically a pillar of trans-Saharan trade for spices, salt and slaves, has once again become a popular trade route, but now the cargo is different. Each week, trucks leave Agadez piled high with passengers who have paid their smugglers to get them to the Libyan coastal city of Tripoli.[27] Many smugglers at the transit posts in Bamako, Ouagadougou and Agadez see themselves as businessmen providing a service. They have the logistics, knowledge of the terrain and ability to bribe officials to look the other way.[28] The police get more money from bribes than their salary, and the demand for the service is growing. People are fleeing Africa to escape slums, unsanitary conditions, ongoing wars and the lack of work, food, water and opportunity. With constant imagery via television and cell phones of what awaits in Europe, young people are willing to risk everything to go north. However, when everything has become a commodity— trees, soil, seed, wildlife, water and nutrients—people become commodities themselves. This is the case for many women and children who are tricked into believing they are getting work in Europe, but actually end up in the sex trade.[29] It is also true for some migrants making their way through Libya to Europe. Some find that in the Wild West of Libyan human smuggling, which is now the fifth largest business in the country, their smuggler can become a trafficker. Some migrants are kidnapped or sold to other smugglers, who in turn torture, rape and imprison them to extract more money from their families.[30] Governments on both sides of the Mediterranean have made concerted efforts to crack down on smuggling with little success because too much money is at stake.[31] The only way to prevent these tragedies is to reduce the need for people to seek out human smugglers by providing security and opportunity at home.

Conflicts and weather-related disasters forcibly displaced 12.4 million people in 2015 alone.[32] While some were able to return home, others found themselves in refugee or internally displaced persons (IDP) camps, and others tried to emigrate. As the largest number of displacements ever recorded, this is equivalent to twenty-four people displaced every minute, and half of them were children.[33] To put it in perspective, imagine by the time you finish reading this chapter you have 180 children outside your door in need of food, water, clothing and shelter. By the time you go to the store, buy supplies and make each of them a sandwich, 1,440 more children have arrived. Does this make it hard for you to breathe? Now imagine you are a parent of one or more of those children, you crossed the Mississippi by canoe and walked across five U.S. states during the winter, only to find a wall of razor wire erected in a field with no supplies or shelter for another one hundred miles.

This is not just a Syrian and European problem. A similar fate awaited the nearly two hundred thousand unaccompanied children who arrived on U.S. soil over the last three years.[34] After fleeing violence and poverty in Central American countries, they walked hundreds of miles, were packed in trains and trucks, endured heat, hunger, thirst, near suffocation and fear only to be shipped to detention centers for processing. To adequately handle another seventy-five thousand unaccompanied children expected to arrive in 2017, the U.S. increased its budget from $200 million in 2013 to $1.3 billion in 2017, a cost of about $17,000 per child.[35] If we are willing to spend $17,000 per child, perhaps we need to reevaluate where we invest this money. Wouldn't it be more productive to invest in the future of a rural child's family farm in Honduras than in processing that traumatized child at a border detention center? For an investment of $640, our forest garden program can help a family develop a sustainable forest garden that provides for about nine people. For the $17,000 cost of processing one child, we can help twenty-six families develop sustainable forest gardens. If each of those families has six children and conservatively two children from each family try to migrate, that is $884,000 that would be saved on processing. That $884,000 could create another 1,381 forest gardens, preventing another 2,722 children from migrating, and so on. So when we look at what it would take to prevent those 75,000 new children from migrating, it begins with a one-time cost of about $24 million. This is a small fraction (only 5 percent) of the budgeted $1.3 billion that involves processing alone. However, that $24 million not only prevents migration, but also provides income, food, hope and a future for 37,500 families, approximately 340,000 people. The opportunities and benefits that flow from these successful family farms include better health, education and prevention of youth from seeking money through criminal activities. In 2012, 90 percent of U.S. foreign assistance for law enforcement and military aid in the region went to countering narcotics.[36] This "foreign assistance" money is only fighting fires, not preventing them. A child does not wake up one morning on a rural farm in Honduras and decide to join the drug trade or walk hundreds of miles to cross the border of another country. It happens over years of hardship, violence, poverty and hopelessness. Looking at my own children I try to image "what if?" What if they woke up one morning and instead of deciding whether to take or buy lunch, they had to decide to whether to join the drug trade or make the long walk? Based on their personalities they would probably walk, but the perils that would await them on this long journey are unimaginable.

The current mass migration from developing countries to Europe and the United States due to poverty, food insecurity, lack of opportunity and violence is only a taste of what's to come. As the global population grows, the climate changes, agriculture fails and violence erupts, people will not die quietly. They will come in masses in search of what we take for granted: food, water, security and opportunity.

The flip side of illegal immigration is the legal immigration that Western countries offer to the most skilled and educated people of developing countries. Although developed countries are enriched by their arrival, the loss of this intellectual resource for their home country is far too great to dismiss. This loss of intellectuals and skilled tradespeople is often referred to as "brain drain." With each loss of an educated, skilled individual, the country loses a teacher, a doctor, an engineer, an electrician, a computer scientist or an agronomist who could have inspired youth, cured illnesses, created infrastructure, started a business or developed an organic farming movement. There are more African engineers working in the United States than in Africa.[37]

In Malawi, there is only one doctor for every fifty thousand people.[38] Educated Africans often leave out of frustration with the lack of infrastructure, supplies and opportunity. Rather than recruiting and facilitating the immigration of highly educated and skilled African professionals, we must invest in infrastructure, salaries and adequate supplies to ensure their success in providing vital services for their populations at home. These kinds of investments are needed for every sector, including agriculture. Rather than exporting corn, wheat, rice, sugar, fertilizers, herbicides and seeds, the U.S. needs to invest in helping African nations sustainably grow and nurture their own natural and human resources. It is not our responsibility to build walls to keep people from migrating. It is our responsibility to grow opportunities, so that migrating is not the only dream young people believe is achievable.

GROWING OPPORTUNITY

In 2007, there was a fundraiser for an organization that collects and sends books to what is now the country of South Sudan. The Sudanese ambassador to the U.S. spoke at the fundraiser. He told how he was one of eleven children born in a small village. He was the only one of his brothers and sisters able to attend school. This struck me. If the only child who went to school became an ambassador, then what could the other ten children have become? What could the other children in his village, state, country and continent have become? How many corrupt regimes thrive on the lack of their population's education? If this generation of children could all attend school, imagine what Africa might become in thirty years.

At Trees for the Future we believe that knowledge is the most powerful thing we can give our farmers. Although most African farmers have very little formal education, they are eager for technical training that can provide tangible benefits without expensive inputs. Rather than subsidizing farmer's inputs, Trees for the Future provides them with knowledge to become masters of their own forest gardens, through our Forest Garden Training Program. Once they master techniques they are certified to teach other farmers and build the strength of the rural farming community. Forest gardens not only grow food and useful products throughout the year, they also grow farmer's income through the sale of diverse locally in-demand products. This profit grows opportunities for their children, including the opportunity to afford to go to school. In addition to growing products, the forest garden grows wood, which for young girls means less time spent searching for firewood and more time in school.

Foreigners have been funding projects in African countries since their independence. Many projects have had limited success because they are developed from the top down. The secret to success is working from the bottom up, building the technical and educational capacity of a country's vast, low-literate, yet extremely resourceful population. Only Africans can innovate, build and grow their countries out of poverty with their own educated and skilled people. Although many well-educated Africans leave for greener pastures in Europe and the United States, others are returning to invest their knowledge in their home countries. Rotimi Williams, a well-educated economist from Nigeria with two masters' degrees and ten years of experience as a journalist in London, is now the second biggest rice farmer in Nigeria.[39] However, Williams isn't just a rice farmer. He uses his rice farm as a business model for community sustainability and peace building. Located in the volatile northern region where Fulani herders and farmers are known to clash, he has created security jobs for men, developed farming activities with women and is instilling a love and knowledge of farming in high school students. His rice is not genetically modified. It is fertilized with organic fertilizer purchased from Fulani herders, and he also provides herders with the rice straw as fodder. All of his rice crop is sold to local markets within Nigeria.[40] This type of closed-loop sustainable farming system is the beginning of the type of agriculture revolution needed in Africa.

At Trees for the Future, as we work to build a sustainable agriculture revolution, we try to get into the minds of farmers and understand what they want through casual focus groups in local languages. One of our focus groups is with young, unmarried men. Gathered in a mud hut with a grass roof, sitting on mats raised off the ground with cinder blocks, these young men wear hand-me-down work clothes and flip flops. Most of them have never set foot inside a school, but they are cheerful, resourceful and know how to work hard. We ask directly why they want to leave the village and migrate to Europe and what it would take to make them stay. Gesturing with rough calloused hands, they describe their needs. Through jokes and serious conversation eventually it comes down to $75 per month. A twenty-two-year-old male would stay in the village and farm if he could make $75 per month, just $2.50 per day. This is the amount of money, the threshold for fulfillment and value, that would make each of them a man in the eyes of their families, setting them on the path to fulfillment. If these young men could achieve that, they would never leave, and ultimately they don't want to leave. Their girlfriends are in the village, their families are in the village. The village is all they know.

When we do focus groups with parents who provide for a family, their desired income is $5 per day. That is only $150 per family per month, less than $2,000 per year. With $5 per day parents can feed their children well, send them to school and patch the roof when it leaks. These are the basic needs that people are unable to meet now. For the millions of twenty-somethings dreaming of $2.50 per day and for the millions of parents dreaming of $5 per day, our forest garden program can get them there and beyond. Through their forest gardens, farmers also get fulfillment—the sky is the limit. Once their gardens are established they can experiment and grow more on their land than they ever thought possible.

My friend Omar who dreamed only of taking a boat north and jumping out of the window of a Red Cross bus to make money in Spain, now makes $3,000 a year with his forest garden and he is finding enormous fulfillment in helping others do the same. Instead of traveling to Europe with nothing, he now travels to neighboring villages and countries with knowledge. He is teaching other farmers how to create sustainable forest gardens. He is helping thousands of families grow opportunities and hope. Now, through the power of trees, he is helping people build houses and futures with money from their forest garden, money they once thought was only attainable in Europe.

By planting millions of these forest gardens we can prevent the need for migration to the slums of cities because farmers can have a good life and support their families on their land. This increased vibrancy in village life also attracts teachers and healthcare workers to remain in or return to the village. By growing hope and opportunities, within families and within the country, forest gardens reduce the desire to migrate north to Europe. Preventing and reducing migration to city slums and other countries significantly reduces the risks of trafficking and recruitment by terrorist organizations. This large-scale change and slowing of rural-to-urban and urban-to-international migration will not happen in one year or five years, but it can happen in one generation. By 2050, in approximately one generation, the planet will be supporting nine billion people and the full effects of climate change will be upon us. This generational shift must happen if we are going to avoid human suffering and mass migration on a scale not yet experienced in the history of humankind. We must start now, not with genetically modified seeds, chemicals and promises, but with tangible action that grows and spreads. With one farmer planting a forest garden and training his or her neighbor to do the same, this movement can spread through the cracks in the industrialized agriculture system. It can spread just as the ideas of intensification and the Green Revolution spread. Only this time it will not be a Green Revolution spearheaded by the Rockefeller Foundation, AGRA or Monsanto. It will be a greener Green Revolution sprouted from the seeds of millions of African farmers dedicated to the simple act of growing forest gardens. However, this endeavor cannot be achieved alone; they need help. Not in the form of technology, but in the form of ideology. If we want to reduce forced migration due to poverty, displacement and war, we must work with these farmers to build opportunities not walls. We must put our energy and investment into their family farms, not detention centers. We must grow forest gardens, not hate.

PART III

ONE WORLD

LOCAL CHANGES WITH GLOBAL IMPACTS

CHAPTER 10

CAN YOU SPARE
SOME CHANGE?

When it is understood that one loses joy and hap-
piness in the attempt to possess them, the essence
of natural farming will be realized. The ultimate
goal of farming is not the growing of crops, but
the cultivation and perfection of human beings.
—MASANOBU FUKUOKA
The One-Straw Revolution

IT STARTS WITH YOUR PIE HOLE

Imagine if all of us, rather than buying the foods that destroy our planet and harm our health, aligned our buying preferences and diets with the wonderful foods that millions of farmers can produce sustainably in forest gardens. This requires us to think about what we buy and where we buy—what we eat, how we eat and where we eat. If we start by looking at our buying and eating habits through the comedic lens of a raccoon, we can more clearly see the ridiculousness of our habits and identify where to change.

Over the Hedge, an animated movie showcasing the insanity of the suburbs was released in 2006. In it a group of forest animals awake from hibernation to find their forest has been transformed into a subdivision. The only thing separating a sea of houses and their remaining plot of forest is a giant hedge.[1] A savvy raccoon, RJ, who has spent years mastering the art of scavenging, begins educating the other animals with the following lessons:

> *"We eat to live, these guys live to eat.*
> *Let me show you what I'm talking about.*
> *The human mouth is called a 'pie hole.'*
> *The human being is called a 'couch potato.'*
> *(Telephone) That is a device to summon food.*
> *(Doorbell) That is one of the many voices of food.*
> *(Front Door) That is the portal for the passing of the food.*
> *(Delivery Truck) That is one of the many food transportation vehicles.*
> *Humans bring the food, take the food, ship the food, they drive the food, they wear the food!*
> *(Microwave) That gets the food hot.*
> *(Cooler) That keeps the food cold.*
> *(Kitchen Table) That is the altar where they worship food.*
> *(Alka-Seltzer) That's what they eat when they eat too much.*
> *(Treadmill) That gets rid of guilt so they can eat more food.*
> *Food! Food! Food! Food! Food!*
> *So you think they have enough?*
> *Well, they don't. For humans, enough is never enough!*

And what do they do with the stuff they don't eat?
They put it in gleaming, silver cans, just for us."[2]

Comedy is the funniest when it gets at core truths, but this truth is so sad that it makes us want to cry. Cry? Or curse? Or change? Dave Deppner, the founder of Trees for the Future, always said, "I plant trees to keep from cursing the night sky." After reading of the many ripple effects modern agriculture is creating throughout the world, you may be asking, *What significant thing can I do to keep from crying or cursing the night sky?* The simplest answer is CWYPIYPH. No, it's not the password for the launch of a nuclear missile. It's an acronym for eight words: Change What You Put in Your Pie Hole. Perhaps it's a bit crude, but you won't forget it and it is much more interesting than the idea of modifying your diet, eating organic or reducing your animal protein intake. The most significant thing you can do to positively impact every problem addressed in this book is to change what you put in your pie hole. Why? Because what you eat drives what farmers grow. Having designed and worked on projects that have reached well over a million smallholder farmers in the last fifteen years, I can tell you definitively that farmers across the world can grow nutritious food in a way that will feed our growing global population while healing the planet. But they simply can't do it without our help. They need us to change our demands. As long as we continue to demand breads, pastas, boxed dinners, crackers, cookies and other junk our doctors tell us not to eat, as well as soy-, palm oil- and corn-based products, farmers will suffer in their pursuit to grow what we want to eat and big ag companies will complete their takeover of the global food system. In the documentary film *Food,*

Inc., Troy Roush, farmer and vice president of the American Corn Growers Association, explained it like this, "You have to understand that we farmers, we're going to deliver to the market place what the market place demands. It's that simple. People have got to start demanding good wholesome food of us and we'll deliver. I promise you. We are very ingenious people."[3] I believe we can make the change by recognizing our habits, understanding where our food comes from and learning how it can be grown in a way that feeds and heals our bodies and planet.

LET US BE EAGLES, NOT CHICKENS

If you have ever watched a chicken eat, it can be quite amusing. It pecks at the corn on the ground directly in front of it with a sort of blind tunnel vision. To get ahead of climate change and the looming food crisis, and we cannot be like chickens (both literally and figuratively) looking at and eating the food we have before us without regard to its impact. By understanding how our food choices are connected to own health and the health of the planet, we can be like eagles. Eagles soar above their food (whether it be a trout or a small rabbit), watching from a distance, with a thoughtful, calculated approach to catching their next meal. Let us be like eagles that soar from the tops of mountains surveying how everything is interwoven below, asking the following questions and making choices based on local, immediate, global and long-term impacts.

☑ **Is it local?**

After digesting the problems outlined in this book, you may feel overwhelmed. In his 2007 *New York Times* article,

"Save the Darfur Puppy," Nicholas Kristof illustrated how the shear enormity of problems makes us shut down and not take action.[4] To to avoid shutting down we must start locally. Buy vegetables and fruits from local organic farmers. This reduces transportation costs, allows you to ask questions about how the product is grown and supports small local farms.

☑ What's in it?

This is the one question all kids have when served something new: "What's in it?" As adults, we stop asking this question, but with processed foods, it's time to start again. Americans do live to eat; we love food. Snack food, fried food, delivered food, fast food, instant food, food trucks, food stands, vending machines, you name it. It is almost impossible to go to an event or anywhere for that matter without finding food. Love of food is not a problem, it's what's in it that is the problem. I get it: we are busy with work and commuting, with school, kids, sports and activities. We don't have time to think about food. We seem to ask only, "Does it taste good? And will it fill me up?" Processed food companies have become very good at fulfilling these two demands. Armed with carbs, salt, sugar and added flavors they have developed irresistible combinations. My personal favorite—Snickers, which is packed with the same monocropped peanuts that are destroying the soils of Senegal. But recently I'm changing my habits so that Snickers becomes a reward I give myself sometimes while my daily snack is a KIND bar because their nuts and fruits are sustainably sourced. Read ingredient labels closely and don't be fooled by labels that say that it contains organic ingredients or is all natural; there is no standard for this labeling. The only organic certification is through the United States Department of Agriculture (USDA) and will contain a USDA Organic seal.[5] One way to ensure that your product is grown by small farmers in forest gardens without chemicals is too look for the Rainforest Alliance Certification. The Rainforest Alliance has a rigorous certification process that ensures the product is grown in a sustainable way for the environment and the farmers. The Rainforest Alliance certifies: bananas, coffee, cacao (chocolate), ferns and cut flowers, palm oil and tea.[6] These foods may be more expensive, but remember: it our choices that drive what farmers produce. To avoid processed food try to ensure half of your grocery cart consists of unpackaged items. What doesn't come in a package? Fruits and vegetables. The very thing our doctors tell us to eat more of to avoid the illnesses plaguing Americans: diabetes, heart disease, high blood pressure and high cholesterol.

☑ **Need or greed?**

Many of the things we buy are impulse purchases or items we buy because we are busy and in a hurry. When we take time to shop we are more conscious about what we buy and more careful to look for ingredients in products. Thus, it is important to make time for food purchasing as well as food preparation. My family eats out about once a week. Each time we spend about an hour reading *Yelp* reviews trying to find the best option. We look at the restaurant atmosphere, ratings, authenticity of the food, menu, kid friendly, etc.

We do not spend an hour each week researching what goes into the food we buy. Prior to purchasing a new phone or computer, think about how much time we spend researching, looking at reviews, technical capabilities, ratings, comparing costs, etc. We are not going to put the phones in our bodies, digest them and count on them to provide us with energy and vitality, yet we spend hours researching them. The loaf of bread, sliced turkey, and tomatoes we buy, on the other hand, will be plucked off the supermarkets shelves without research or thought of the high fructose genetically modified corn syrup, growth hormones or insecticides they may contain. But the thing is, we must eat. We want to eat. Food is pleasure. So how do we eat without killing ourselves? Before buying anything, ask yourself, "Was this grown or produced for need or greed?" If you answer greed, don't buy it. Sometimes the line between the two is grey, but often the choice is clear. In the supermarket there are many clear choices: apples (need), apple flavored fruit rollups (greed), baguette bread (need), powdered mini donuts (greed), and so on. When you view it immediately, two apples are about the cost of a whole box of fruit rollups. However, when weighing the health benefits, the apples cost less. The items produced for greed contain derivatives of corn, soy, palm oil or sugar—all things our doctors tell us to avoid eating too much of. Healthy foods cannot be sustainably produced at extremely low prices and still provide a livelihood for farmers. The need or greed test can be expanded to other parts of your life, including the numerous cheap plastic birthday party favors and other wasteful products that are bought to fulfill an immediate impulse and quickly go from the store to the trashcan.

☑ What do I want to be made of?

The two big things that affect large U.S. farmers, small developing-world farmers and livestock owners everywhere are corn and meat. By changing our consumption of these two things we can drastically impact agriculture.

Soon after graduating from college, documentary film makers Ian Cheney and Curtis Ellis discovered that for the first time in American history, their generation was at risk of having a shorter lifespan than their parents due to what they ate. Active and healthy, they couldn't understand what they were eating that was so detrimental. Learning that their hair acts as a tape recorder of their diet, they had an isotope analysis done on their hair. The analysis revealed that the majority of the carbon in their bodies originated from corn. Ian and Curtis were astonished by these results because they rarely ate corn, but the corn in their hair wasn't from corn on the cob. It was from invisible corn: corn sweeteners, corn starch and corn derived products in the processed foods and drinks they had eaten from an early age. It was also from corn-fed animals, their primary sources of protein. Nearly everything they ate was corn: meat, spaghetti sauce, bread, apple juice, etc. Not only was their diet made of corn, but genetically modified (GM) corn.[7]

Jamais Cascio, a writer, futurist and contributor to the documentary film *Six Degrees Could Change the World*, recently calculated the combined carbon footprint of Americans eating cheeseburgers in one year and found it to be two-hundred million metric tons (producing more CO_2 than all the SUVs in the U.S.). The majority of that footprint comes from cattle and the methane they produce.[8] In addition, more than two-thirds of all agricultural land is devoted to growing feed for livestock, while only 8 percent is used to grow food for direct human consumption. Small farms with free-roaming animals are disappearing in many parts of the world. Currently, three-quarters of the world's poultry supply, half of the pork and two-thirds of the eggs come from industrial meat factories.[9] The concentration of livestock increases the environmental burden. Reducing the amount of meat we consume by one to two times a week can have a significant impact, not only on greenhouse gas emissions, but also on our health. A recent study at Harvard University found that for each additional serving (3 oz.) of red meat our "risk of mortality increases by 13% if the meat is unprocessed and 20% if it is processed."[10]

☑ What is being learned from this?

In the documentary film *Before the Flood*, Leonardo DiCaprio sits down with Dr. Sunita Narain of the Center for Science and Environment in Delhi, India, to talk about climate change.[11] Although her strong comments refer to energy efficiency and renewable energy, they can be applied to our agriculture systems and food supply as well. She posed the question, "What is the United States doing which the rest of the world can learn from? You are a fossil-addicted country, but if you are seriously disengaging, that's something for us to learn from. And it's leadership that we can hold up to our government and say if the US is doing it—then, despite all the pressures, then we can do it as well . . . But it's just not happening . . ."[12] In this statement Dr. Narain suggests that if we disengaged from the use of fossil fuels it would serve as an example to the world. However, we are not disengaging from fossil fuels, rather we are disengaging from the climate change conversation and our role in solving it. Is this what we want the world to learn from us, disengagement from responsibility? This is not the type of leadership she was implying. The world emulates our actions—what we buy, what we do and how we do it. If we can make these necessary changes in food choices and agriculture systems, the rest of the world will do it too. Around the world people want to eat McDonalds because we do. As Americans we have an opportunity to export our newfound belief in *grow organic, buy local, eat veggies,* and *meatless Mondays* to millions of small-scale rural farmers in developing countries. If we say no to the thousands of genetically modified corn- and soy-infused processed foods that line our supermarket shelves, then we are saying yes to small organic farmers, and no to industrialized crops. This momentum needs your help to sustain it. We may not realize that the decisions we make on a daily basis have repercussions on Mariama in Senegal, cattle in the Sahel, Limbi in Cameroon and soil in Ethiopia, but they do. The world mimics our actions. By saying no to processed foods and plantation crops, we are validating the choices that small farmers are making. We are saying, "Yes, you got it right, keep it up." You are leading the food systems of the future. Our choices, like the roots of a tree, will spread, nourishing and stabilizing small farmers, giving them hope and independence. And with this independence more food can be grown in gardens that heal the earth and fortify our bodies.

☑ What will next season look like?

There was once a man who had four sons. He asked each of them to go to the distant valley and describe the large tree they found there. He sent the first son in winter, the second in spring, the third in summer and the fourth in the fall. After they each had their turn, he gathered them together. "Tell me, my sons," he said, "what did you see?" The first son described a dead tree with barren branches covered in ice. The second son told of a tree alive with bright flowers. The third spoke of a tree thick with bright green leaves and teeming with song birds, and the fourth described a golden tree laden with juicy fruit. Each of the brothers thought the others were mad. How could they have each seen something so vastly different? "You are all right," said their father. "The tree, like people and life, has many seasons. We must experience them all before we can know what gifts they have to offer."[13] Taking this lesson from a tree, we must not judge our current state of agriculture as the only season of the future. Right now we are experiencing the winter of modern day agriculture. However, by changing our purchasing and eating habits we can begin to enter a new season characterized by abundance.

Local and immediate individual choices have global and long-term impacts. By making many small choices each day, we can drive the creation of millions of forest gardens, leaving our legacy on the earth, a living legacy for generations. This can be a clean, green legacy where fruits, vegetables and nuts are plentiful, where trees are respected because we know they give us all that we need: food, shade, air, water, security, hope, memories and a life worth living. By changing what you put in your pie hole, you change your health, your weight, farming practices and the state of our local and global environments in one shot.

MAKE AGRICULTURE A PART OF YOUR CULTURE

A century ago in the United States, farmers made up nearly a third of the labor force. Today they are just over 1 percent.[14] Why? Because growing food is hard work and it is not very lucrative. Under the agriculture systems we have in place, it is easy to work very hard and fail. No one says, "I want to grow up and be a farmer so I can get rich." Even when farmers can manage their debts and keep their head above water, they often feel disappointment. Farmers chose their occupation out of the need to produce and grow food. When crops fail due to weather or pests, U.S. farmers have insurance to cover part of the financial loss, but it does not cover the emotional debt of losing one year's worth of work. The knowledge that they didn't produce anything for people to eat weighs heavily. We can begin to understand the plight of farmers by stepping into their shoes. One of the most impactful ways to begin to understand our food, where it comes from and what it takes to grow it, is to try growing it ourselves. You can start small with a couple vegetables and herbs and expand from there. There are numerous ways in which you can do this. If you own a house with a backyard, start there. If you don't, find one. The website www.sharedearth.com connects nearby landowners who want gardening on their property to people who want to garden. Likewise, towns and cities throughout the United States offer community gardening opportunities. For about $100 per year you can rent a gardening plot and produce some of your own vegetables. Make it a family activity. Kids love to play in the dirt, plant seeds, watch them come up and learn what different vegetables look like at different stages. They also love to harvest, cook and eat the food they grow. There is not only great pleasure in growing your own vegetables, but also great taste. With just a small plot you can grow much more than the cost of the rental fee. In fact you will probably grow more than you can consume, giving you the pleasure of sharing the food you grow with friends and neighbors.

Another way to start making agriculture a part of your culture is to get involved with Community Supported Agriculture (CSA) groups. You can find CSAs in many cities throughout the United States. There are several different fee structures. Some request participants pay up front prior to the start of the growing season so that farmers have capital to operate without borrowing money.[15] Others have a monthly price structure. By joining a CSA you can have fresh, local vegetables and fruits delivered to your door on a weekly or biweekly basis. This is a great way to support local agriculture and learn to eat and cook with vegetables that are in season. If you are on a tight budget, in exchange for volunteering to pack and deliver boxes, you can often get your vegetables free of charge.

By understanding both the difficulties and the love of producing food, we can begin to understand what farmers experience, see them as our heroes and become invested in their success. It is our duty to not only change what we put in our pie hole, but support the farmers who work hard to bring us healthy food and encourage the farmers who are stuck in the cycle of monocropping to transition to more sustainable agriculture practices.

As referenced in the introduction to this book, monocrop farming gives farmers about forty chances in their lifetime to get it right.[16] Forty chances is not much. For many of us we have a chance every day or week or month to prove ourselves and accomplish goals. What if we only had one chance a year? As we quickly deplete our resources, we are down to one shot to get our food systems right. If we only have one shot this century, where will we put our chance for change—in a genetically modified corn field or a forest garden?

CHAPTER 11

THE GUILDED AGE

Cultivators of the earth are the most valuable
citizens. They are the most vigorous, the most
independent, the most virtuous, and they are tied
to their country and webbed to its liberty and
interests by the most lasting bonds.
— THOMAS JEFFERSON

METHODOLOGY NOT TECHNOLOGY— GROWING GUILDS

In the summer of 2015, Bill Gates launched a video entitled *Who Will Suffer Most from Climate Change? (Hint: Not You)*. The video was meant to show how small farmers can improve agriculture output and in turn their lives. However, the prescription called for better seeds and chemical fertilizer.[1] Coming from someone who built their career on innovation, it makes sense that Bill Gates looks to technology for solutions. However, technology cannot fix farmers' most pressing problems. Our ability to sustainably feed the world with adequate nutrients in the coming decades lies in focusing on methodology, not technology.

A friend of mine recently told me a story that illustrates this point so well. Her colleague's father is a professor of agricultural genetics at a university in Sudan, north of the capital Khartoum. He has been working for years to develop hybrid seeds that are drought tolerant and produce greater yields. He and his associates test these seeds on a large research plot at the university, adjacent to the animal husbandry and veterinary research buildings. One weekend, someone who was looking after the animals let the cows into the agriculture plot—and that was the end of the seed tests for an entire year. Methodology, not technology. Farmers in Africa can be given the best genetically developed seeds that science can create. These seeds can be planted with the most appropriate technology for the region. However, when these sprouted seeds are eaten by large herds of cows, camels, sheep and goats that roam freely over the continent, the time, money and effort is lost.

While writing this book, I spent a week in Malawi at the Kusamala Institute for Agriculture and Ecology training trainers across four large organizations to replicate our forest garden approach. Posted on a large tree in the institute, right next to the permagarden demonstration area, a sign reads:

The Guild

A 'guild' is a term in Permaculture that is

used to describe plants and animals that

work together to support each other.

A guild should have:

Food for us

Food for the soil

Groundcovers

Diggers

Protectors

Climbers

Supporters

A guild is designed to mimic what happens

in a forest, while at the same time

providing us with foods, medicines,

building supplies, fuels, fiber, and more.

No chemicals. No deforestation. No need for the world's poorest people to send 60 percent of their profits to chemical and seed suppliers and pest control providers. No burning. No clearing. No wiping out of forests and biodiversity to feed ourselves. No advancing desertification. Less risk. Better food and a lot more of it.

There is no silver bullet to sustainable agriculture production. There is no cookie-cutter plan but rather a methodology that involves a systemic approach to providing for the land, the farmers and the community. It involves the development of agroforestry farms, forest gardens, guilds and permacultures that combine the best of a forest and a garden in one place. These systems mimic the sustainability of a natural forest by incorporating a variety of tree species and other crops in a horizontal and vertical design, resulting in sustainable sources of food, firewood (energy), fodder, natural fertilizer, natural insect control and diverse marketable products that provide income throughout the year. A permaculture is a permanence of agriculture, one that does not end with each harvest and begin again from scratch but rather creates a continuous cycle of new opportunities for growth and experimentation within a thriving ecosystem.

Remember Keba Mbengue, my friend and farmer from the introduction to this book? Keba created this permanence of agriculture. Keba used to put a lot of grains in his pie hole. He used to grow monocrops of peanuts, millet and maize. He would give his money to the fertilizer salesmen who were happy to take it and bring him more each year as his fields degraded over time, making him more dependent upon them. He understood his way of farming was not providing for him, his family or his community. He realized that there was a local demand for fruits and vegetables that weren't being grown. When Keba started planting a forest garden he didn't do it to change the global food system; he did it to lift his family out of poverty, and in the process he replicated a model that has the potential to end global poverty, hunger, climate change and the loss of the world's forests and biodiversity. Keba turned an acre that once produced only $200 a year into a forest garden system that generates a 500-percent increase (nearly $1000) in revenue every year. Each year his forest garden produces more and more money with less and less labor.

On a one-acre plot that once grew cash crops for five months and sat empty and barren for seven months, Keba now has a forest garden with a couple thousand trees. He made the change from a one layer cash crop to a five layer forest garden. When asked if he would ever go back to peanuts, he shook his head and said, "I believe in trees." Keba helped me believe in both trees and people's ability to change their lives through them.

KEBA'S FARM INCOME EACH YEAR	
BEFORE the Forest Garden	**AFTER** the Forest Garden
$200 from cash crop once a year	$400 in cashew nuts
	$250 in cashew apples—that he gives away to village children
	$80 in jujubes (desert apples)
	$230 from eggplant, hot peppers and other vegetables intercropped among trees
	100% of his own firewood supply
	Medicinal and nutrient-providing leaves from acacia and moringa trees
	Compost and natural fertilizers from trees

As the world quickly approaches nine billion people, a philosophical question I ponder over wine with friends is, with technology and automation reducing the need for labor, what are all these billions of people going to do to find fulfillment? The current farming system is so painful that youth are fleeing rural areas to escape the grasp of farm debt, climate change and desertification. But farming the land through a forest garden, filled with interesting and lucrative diversity, provides an opportunity for us to give the next generation the fulfillment from farming that has been lost. As the land grows and diversifies, the farmer further innovates, and as the land reaches its full potential, so does the farmer. Together they develop what nature intended: a symbiotic relationship that can't be achieved through technology.

WHERE TO PUT YOUR BENJAMINS—A CHECKLIST FOR SUSTAINABLE FOOD PROGRAMS

Throughout this book, we've learned of the impact agriculture has on people, the planet and the farmer's profit. We've learned of the personal ways we can make food changes in our lives that ultimately impact the future of agriculture. But how and where can we invest our time and Benjamins (money) to have a direct impact on agriculture and the future of food? As donors, scientists, directors, planners and field technicians working in sustainable development we must ask ourselves, "Does the agriculture system we are supporting continuously make things better for people, the planet and the farmer's profit? If successful, is this the solution that the world needs now and in the future? Who stands to benefit most from this solution? Does this solution reflect what farmers want?" If the solution we are developing or supporting does not correctly answer these questions, then we need to stop and re-evaluate. Agriculture programs that do not meet these standards are socially and environmentally setting us back. The following is a checklist identifying the effective features of sustainable agriculture programs.

☑ **The program meets a triple bottom line positively impacting People, the Planet and Farmer's Profit.**

The first question involves a triple bottom line—PPP (people, planet, profit). **People** in poor rural areas of developing countries have basic needs that must be met by the program, including diverse foods, fuelwood and fodder for their animals.

TRIPLE BOTTOM LINE	SUSTAINABLE AGRICULTURE SYSTEMS MUST
PEOPLE	Be permanent and continually make things better
	Protect people from hardships (drought, flood, storms)
	Minimize risk—farmers have so many risks already; this needs to be a sure thing
PLANET	Lead to sustainability, provide resources (wood/energy, animal forage, food, commodities to sell) and alleviate pressures on the land
	Help the land continually grow and produce; provide for people while at the same time regenerating and building soils
PROFIT	Provide income to afford necessities that can't be grown: end the cycle of poverty and farm debt
	Provide income throughout the year instead of one payday
	Show a return on investment: not only more income, but also more time, more resources, more opportunity

For the land to have the capacity to produce these things, the program must protect the **planet** by continuously caring for and improving the quality of land, air, water and other natural resources. Finally, the program must provide a **profit** for people to afford other essentials: housing, clothing and schooling for their children. There are numerous agriculture programs aimed at improving people's lives by increasing yields, accessing better markets, microfinancing and provision of inputs. Although many programs address one or two of the aspects of the triple bottom line, very few address all three. Therefore, the first question is, does the solution meet the PPP checklist?

☑ **The program's success is what we need now and in the future.**

It is important for us to look at the future impact of a program by asking ourselves, what happens if this program is implemented, everyone adopts it and it becomes wildly successful? Is this what the world needs now and in the future? In 1953, in response to deteriorating soils, failing crops and hunger in Mexico, Dr. Norman Borlaug worked tirelessly to develop a wheat variety that with the application of large amounts of chemical fertilizer, could triple yields. He did this by developing a dwarfed stalk variety, meaning short, thick stalks that could support the growth of large wheat heads without toppling over. Many farmers across Mexico soon adopted this practice and by the early 1960s Mexico's wheat harvest increased six-fold. This practice quickly spread to India, where a similar food crisis existed.[2] By 1968 the Indian wheat harvest "was so bountiful that the government had to turn schools into temporarily granaries."[3] Later this technology was adapted to rice varieties in the Philippines and throughout Asia. Dr. Borlaug won the Noble Peace Prize in 1970 and was labeled as the father of the first Green Revolution in Latin American and Asia.[4] However, decades later, this Green Revolution turned everything brown, subjecting the children and grandchildren of those farmers to extreme poverty, hunger and deprivation. In trying to solve the hunger crisis of his time, Dr. Borlaug unknowingly created a grain dominance within the world's food supply, "displaced smaller farmers, encouraged overreliance on chemicals and paved the way for greater corporate control of agriculture."[5] In her book *The Violence of the Green Revolution*, Dr. Vandana Shiva wrote that the Green Revolution strategy ". . . differed from the indigenous strategies primarily in lack of respect for nature's processes and people's knowledge. In mistakenly identifying the sustainable and lasting as backward and primitive, and in perceiving nature's limits as constraints on productivity that had to be removed, American experts spread ecologically destructive and unsustainable agricultural practices worldwide."[6] We do not want future agriculture programs to repeat this history; thus the solution must listen to the farmer and nature, and its success must improve life today and into the future.

☑ **The farmer is the underlying beneficiary.**

The third question revolves around who stands to benefit the most from this solution. If our answer is "everyone but the farmers," then the strategy must be reconsidered. For example, there are government agencies and organizations working in Africa that provide subsidies, microfinancing and layaway plans for farmers to get improved seeds, chemical fertilizer and insecticide packages. These inputs promise to increase yields and double incomes. Although these programs may meet the profit aspect of PPP, they fail to meet the people and planet aspects. Why? Because increasing yields on a monocrop of peanuts or maize or any other crop will only provide more of that crop to sell at a time when prices are the lowest, meaning a very modest increase in income once a year. An increased harvest of peanuts or corn will not meet the nutritional needs of a family nor provide a multitude of diverse products for people to sell and use throughout the year. The chemical fertilizers poured on the land will not leave lasting nutrients in the soil; once the crop is harvested, the nutrients are gone. Insecticides and herbicides poison the land, water and the biodiversity that the land needs to thrive. Thus, when we look at who the underlying beneficiary is, we do not find the farmer but rather the producers of seeds, fertilizers and insecticides.

☑ **The program is what the farmers want.**

When working with farmers we must ask, "What do you want to grow?" If micro-lending institutions polled the farmers in their programs, they may be surprised to find that their farmers are saving to plant and fertilize crops they no longer want to grow. Humans are not easily satisfied with being stagnant. Being poor does not change our desire to strive for something more. In the development community, we have entered a neocolonial situation where the chains are still there through the presumption that people want more of what they have. People who live under a tarp are not looking for a bigger tarp. They are striving to build a dwelling that doesn't leak and protects them from storms and insects. Likewise, peanut farmers who do not produce anything they can eat are not looking to save their money to purchase expensive fertilizer that provides a slight increase in yields but doesn't improve their family's nutrition, enhance the quality of their land and change the course of their future. They are looking for the fulfillment that comes from producing bountiful, healthy food for their families and communities. The fulfillment that comes from rising early to see the budding of fruit trees. The fulfillment that comes from hearing what their children learned at school that day.

With this in mind, this program must first consider a farmer's vision for their future. For cash and monocrop farmers, every year is like the 1993 movie *Ground Hog Day*,[7] the nightmare of starting over, with a bare field, no seeds, no nutrients and very little soil moisture, repeating the same process with no positive change in outcome. Although farmers have one of the most value-added products—transforming seeds into edible foods—they get the least return on their investment. Farming is perhaps the only business in the world where owners buy at retail (inputs of seeds, fertilizer and insecticides) and sell at wholesale (grains and cash crops). This is opposite of most businesses, which buy their base ingredients at wholesale and sell their product at retail. Knowing this, we must help farmers develop an agriculture system that builds capital from year to year, that creates numerous streams of revenue, that continuously grows and provides internal cycles of nutrients.

☑ **The program's success is measured in a logical way.**

The program must have a method for logically measuring success. If I decide to lose weight, how would I measure my success? Maybe I could count the number of "low fat" products I eat each day. Or perhaps I could count the number of meals I skip. Although these metrics might look impressive on a chart, their validity in measuring successful weight loss is ludicrous. Likewise, we must be mindful of how "sustainable agriculture" programs are being measured. When we look at monocrop subsistence farmers in rural parts of the developing world, their grain yields are often so insignificant to the global market that they could be equivalent to what falls on the ground during U.S. harvest seasons.[8,9] Although their outputs are meager, their external inputs (seeds, fertilizer and insecticides) are significant.[10] In fact, it might seem that the only reason to encourage smallholder farmers to produce peanuts, corn, soybeans or cotton is to promote the purchase of inputs (seeds, fertilizer and insecticide). Thus, some programs are using the quantity of seed and fertilizer purchased as a metric for measuring success. Just as the number of "low fat" products I eat does not measure successful weight loss, the amount of fertilizer purchased does not measure successful crop production, and it certainly is not a measure of sustainable crop production.

Whether you are donating, working or volunteering for an agricultural program, this checklist can help you determine the most appropriate way to use your time and Benjamins to achieve a food system that will help us all thrive.

PRO-VICTION NOT CON-VICTION

"From Little Things Big Things Grow," a folk song by Australian artist Paul Kelly and Kev Carmody, tells the inspiring true story of Vincent Lingari's fight for his people's native land. In the late 1800s British pastoralists forced the Gurindji people off their lands and killed anyone who protested. The Vestey Brothers, a large food production company, purchased the land in 1914 and used forced Gurindji slave labor to work and expand their farmlands. While respect for human rights took root in the rest of the country, due to the isolation of the land, this forced labor continued for another fifty years. One hot day in August, 1966, Vincent Lingari began a strike and other men followed. The Vesteys offered the Gurindji a small salary and better living conditions, but they knew what they wanted. They wanted their land back. The Gurindji people continued their strike while Vincent traveled the country telling his story. Eight years after Vincent began his strike, on August 16, 1975, Prime Minister Whitley gave the Gurindji the deed to their land and "through Vincent's fingers, poured a handful of sand," representing the return of the land to his people.[11] Vincent's story is a powerful tale of what one man can achieve when he stands up for what he wants. In our work in Africa we have found that farmers do not want the modern-day version of what the Vesteys were offering (subsidized seeds and fertilizers) that will only bring them greater yields but no real change in their future opportunities. Farmers want what Vincent wanted, sovereignty over their land and what they produce. They want an opportunity to produce more than substance, to grow what they need and use what they grow. Farming through the forest garden gives them that.

Life coach Heidi Dewan could be considered a farmer of souls. She helps people nurture and grow their inner wisdom and confidence to succeed. Her simple yet powerful message tells us, "Whatever we focus on gains power. Our efforts for positive change are powerful when we rally around something that we want, not against something we do not. Not only can our energy go towards creating beauty, goodness and love in the world, no matter how small and localized, but we are also practicing building something up, rather than tearing it down. Many governments that replaced their corrupt predecessors ended up being a disappointment to the people. Why? Because the revolution was focused on fighting against injustice and the institutions that enforced it."[12] We know that we do not want exploitation of the land and people, lies, greed, deceit, hunger and poverty. We cannot fight Monsanto or Dupont, we cannot fight Tyson Foods or Nabisco but we can take this lesson to heart and focus on what we want for people and the planet.

What do we want?

Healthy food and communities. Unity. Human rights. Truth. Protection of our earth and respect for soil, water and all creatures. This is created through the practice of building something up, through the growth of millions of forest gardens.

END NOTE REFERENCES

INTRODUCTION: A TALE OF FIVE FATHERS

1. Richard J. Johnson and Timothy Gower, The Sugar Fix: The High-Fructose Fallout That Is Making You Fat and Sick (New York: Pocket Books, 2009).
2. Joseph Mercola, "9 Ways That Eating Processed Food Made the World Sick and Fat," Mercola, February 12, 2014, http://articles.mercola.com/sites/articles/archive/2014/02/12/9-dangers-processed-foods.aspx.
3. Howard G. Buffett, 40 Chances: Finding Hope in a Hungry World (New York: Simon and Schuster, 2013), 10.
4. Buffett, 40 Chances: Finding Hope in a Hungry World, 10.
5. Ivette Perfecto, John H. Vandermeer, and Angus Lindsay Wright, Nature's Matrix: Linking Agriculture, Conservation and Food Sovereignty (London: Earthscan, 2009).

CHAPTER 1: YESTERDAY'S PROMISE

1. Kevin Shillington, History of Africa, 2nd ed. (New York: Macmillan, 2005).

2. "Small Holders and Family Farmers," Factsheet, Food and Agricultural Organization of the United Nations, Sustainability Pathways, 2012.

3. "Cattle Ranching and Deforestation," Livestock Policy Brief 3, Livestock Information, Sector Analysis and Policy Branch Animal Production and Health Division (Rome: Food and Agricultural Organization of the United Nations, 2006), http://www.fao.org/3/a-a0262e.pdf.

4. Betty (corn farmer), in discussion with Brandy Lellou, February 15, 2014.

5. United States Department of Agriculture, Grain: World Markets and Trade (Washington, DC: Foreign Agriculture Service, United States Department of Agriculture, December 2016), https://apps.fas.usda.gov/psdonline/circulars/grain-corn-coarsegrains.pdf.

6. Ralph E. Nield and James E. Newman, "Growing Season Requirements and Characteristics in the Corn Belt," in *The National Corn Handbook*, https://www.extension.purdue.edu/extmedia/nch/nch-40.html.

7. Betty (corn farmer), in discussion with Brandy Lellou, February 15, 2014.

8. "Corn Breeding: Lessons From the Past," Department of Agronomy and Horticulture, University of Nebraska—Lincoln, accessed February 10, 2017, http://passel.unl.edu/pages/informationmodule.php?idinformationmodule=1075412493&topicorder=3&maxto=12.

9. R. L. Nielsen, "Drought and Heat Stress Effects on Corn Pollination," Department of Agronomy, Purdue University, accessed February 10, 2017, https://www.agry.purdue.edu/ext/corn/pubs/corn-07.htm.

10. Betty (corn farmer), in discussion with Brandy Lellou, February 15, 2014.

11. Patti Domm, "Massive US Drought Leads to Worst Fears for Corn Crop," *Market Insider*, CNBC, August 10, 2012, http://www.cnbc.com/id/48610120.

12. Betty (corn farmer), in discussion with Brandy Lellou, February 15, 2014.

13. Laurel Bower Burgmaier, *The Farm Crisis*, film, produced by Iowa Public Television, July 1, 2013, http://site.iptv.org/video/story/2388/the-farm-crisis.

14. Shillington, *History of Africa*.

15. Pierre Montagne and Oumarou Amadou, "Rural Districts and Community Forest Management and the Fight against Poverty in Niger," *Field Actions Science Reports*, Special Issue 6, (2012), https://factsreports.revues.org/1473.

16. Shillington, *History of Africa*.

17. Justin Gillis, "Norman Borlaug, Plant Scientist Who Fought Famine, Dies at 95," *The New York Times*, September 13, 2009, http://www.nytimes.com/2009/09/14/business/energy-environment/14borlaug.html.

18. Shillington, *History of Africa*.

19. John Steinbeck, *Grapes of Wrath*, reprint (New York: Penguin Group, 2006).

20. United Nations Department of Economic and Social Affairs, *World Urbanization Prospects: The 2014 Revision; Highlights* (New York: United Nations, 2014), available from https://esa.un.org/unpd/wup/publications/files/wup2014-highlights.pdf.

21. United Nations Office on Drugs and Crime, *Global Report on Trafficking in Persons*, (Vienna: United Nations, 2014),

http://www.unodc.org/documents/data-and-analysis/glotip/ GLOTIP_2014_full_report.pdf.

22. United Nations High Commissioner on Refugees, *Global Trends: Forced Displacement in 2015* (Geneva: UNHCR, 2016), available from http://www.unhcr.org/576408cd7.pdf.

23. Bower Burgmaier, *The Farm Crisis.*

24. Eva Orsmond, "Sugar Crash," YouTube video, 51:08, from a documentary televised by RTE Television on January 11, 2016, posted by "EL MUNDO Y SUS CREACIONES," November 23, 2016, https://www.youtube.com/watch?v=_ CNcVW-Uo6Q&feature=youtu.be

25. Bower Burgmaier, *The Farm Crisis.*

26. Joel Achenbach and Dan Keating, "A new divide in American death," *The Washington Post,* April 10, 2016, http://www. washingtonpost.com/sf/national/2016/04/10/a-new-divide-in-american-death/?hpid=hp_no-name_whitedeath-underdis-play_1%3Ahomepage%2Fstory&tid=a_inl.

27. Max Kutner, "Death on the Farm," *Newsweek,* April 10, 2016, http://www.newsweek.com/2014/04/18/death-farm-248127.html.

28. Alice Park, "You Won't Believe How Much Processed Food Americans Eat," *Time*, March 9, 2016, http://time. com/4252515/calories-processed-food/.

CHAPTER 2: TOMORROW'S UNCERTAINTY

1. Micucci Charles, *The Life and Times of Corn* (New York: Houghton Mifflin Harcourt Publishing Company, 2009).

2. Charles, *The Life and Times of Corn.*

3. Jorge Fernandez-Cornejo, Seth Wechsler, Mike Livingston, and Lorraine Mitchell, General Assembly resolution 70/1,

Transforming Our World: the 2030 Agenda for Sustainable Development, A/RES/70/1 (25 September 2015), available from https://sustainabledevelopment.un.org/post2015/transformingourworld.

4. United Nations, Department of Public Information, *Global Goals campaign announced with UNDP as key partner. 3 September 2015*, http://www.undp.org/content/undp/en/home/presscenter/pressreleases/2015/09/03/global-goals-campaign-2015.html.

5. Vararat Atisophon, Jesus Bueren, Gregory De Paepe, Christopher Garroway, and Jean-Philippe Stijns, "Revisiting MDG Cost Estimates From A Domestic Resource Mobilisation Perspective," (working paper no. 306, OECD Development Centre, Paris, France, December 11).

6. Suzanne Simard, "How Trees Talk to Each Other," filmed June 2016, TED video, 18:19, https://www.ted.com/talks/suzanne_simard_how_trees_talk_to_each_other.

7. Dan McKinney, "Trees Found to Communicate through Fungi [Avatar!]," YouTube video, 4:40, from the film *Do Trees Communicate*, produced by Julia Dordel, 2011, posted by "MushrooMetropolis," November 23, 2014, https://www.youtube.com/watch?v=AJWE3y2xdhQ&-feature=youtu.be

8. "Current World Population," Worldometers, owned by Dadax, accessed October 20, 2017, http://www.worldometers.info/world-population/.

9. Lawrence Haddad, Corinna Hawkes, Emorn Udomkesmalee, Endang Achadi, Mohamed Ag Bendech, Arti Ahuja, Zulfiqar Bhutta et al., *Global Nutrition Report 2016: From Promise*

to Impact; Ending Malnutrition by 2030 (Washington, DC: International Food Policy Research Institute, 2016).

10. Frances E. Aboud and Aisha K Yousafzai, "Very Early Childhood Development," in *Reproductive, Maternal, Newborn, and Child Health: Disease Control Priorities, Third Edition*, Vol. 2, ed. R. E. Black, R. Laxminarayan, M. Temmerman, et al. (Washington, DC: The International Bank for Reconstruction and Development/The World Bank, 2016), https://www.ncbi.nlm.nih.gov/books/NBK361924/.

11. "Collecting Water in Ethiopia," Water.org, accessed July 7, 2017, http://water.org/post-v/collecting-water-ethiopia/.

12. Lester R. Brown and Earth Policy Institute. *World on the Edge: How to Prevent Environmental and Economic Collapse.* New York: W.W. Norton, 2011.

13. Baba Umar, "India's shocking farmer suicide epidemic," *Aljazeera,* May 18, 2015, http://www.aljazeera.com/indepth/features/2015/05/india-shocking-farmer-suicide-epidemic-150513121717412.html.

14. Oliver Milman, "Earth Has Lost a Third of Arable Land in Past 40 Years, Scientists Say," *The Guardian,* December 2, 2015, https://www.theguardian.com/environment/2015/dec/02/arable-land-soil-food-security-shortage.

15. "Haiti Statistics," Haiti Partners, last modified November 18, 2015, https://haitipartners.org/about-us/haiti-statistics/.

16. "Conference Puts Focus on Human Trafficking, Fastest Growing Criminal Industry" (Newspaper article, William Spindler, United Nations High Commissioner for Refugees, Lille, 11 October 2010), http://www.unhcr.org/en-us/news/latest/2010/10/4cb315c96/confer-

ence-puts-focus-human-trafficking-fastest-growing-criminal-industry.html.

17. Michaeleen Doucleff, "Anthrax Outbreak in Russia Thought To Be Result of Thawing Permafrost," *Morning Edition*, August 3, 2016, transcript and radio broadcast, 3:34, http://www.npr.org/sections/goatsandsoda/2016/08/03/488400947/anthrax-outbreak-in-russia-thought-to-be-result-of-thawing-permafrost.

18. IPCC, 2013: Summary for Policymakers. In: *Climate Change 2013: The Physical Science Basis*. Contribution of Working Group I to the Fifth Assessment Report of the Intergovernmental Panel on Climate Change [Stocker, T.F., D. Qin, G.-K. Plattner, M. Tignor, S.K. Allen, J. Boschung, A. Nauels, Y. Xia, V. Bex and P.M. Midgley (eds.)]. Cambridge University Press, Cambridge, United Kingdom and New York, NY, USA.

19. "Goal 2: End Hunger, Achieve Food Security and Improved Nutrition and Promote Sustainable Agriculture" (Sustainable Development Goals, United Nations, 1 January 2016), http://www.un.org/sustainabledevelopment/hunger/.

20. Clayton M. Christensen, Michael E. Raynor, and Rory McDonald, "What Is Disruptive Innovation?" *Harvard Business Review*, December 2015, https://hbr.org/2015/12/what-is-disruptive-innovation.

21. "Population Density India," MapsOfIndia.com, last modified August 25, 2011, http://www.mapsofindia.com/census2011/population-density.html.

22. Eric Toensmeier, *The Carbon Farming Solution* (White River Junction: Chelsea Green Publishing, 2016).

23. M. Abdul Salam, K. Sathees Babu, and N. Mohanakumaran, "Home Garden Agriculture in Kerala Revisited," *Food and Nutrition Bulletin* (United Nations University, Tokyo) 16, no. 3 (1995), http://archive.unu.edu/unupress/food/8F163e/8F163E07.htm.

CHAPTER 3: HOOVES AND AXES BREAK THE LAND

1. Myra Busby, "Along the Great Western Cattle Trail," Seymour Chamber of Commerce, accessed September 7, 2016. http://www.greatwesterncattletrail.com/along_gwct_a/along_gwct.html.

2. Garrett Hardin, "Tragedy of the Commons," in *The Fortune Encyclopedia of Economics* ed. David R. Henderson (New York: Warner Books, 1993), http://oll.libertyfund.org/titles/1064.

3. "The Dust Bowl: Legacy," Ken Burns, 2012, http://www.pbs.org/kenburns/dustbowl/legacy/.

4. "Table 1. Urban and Rural Population: 1900 to 1990," U.S. Census Bureau, October 1995, https://www.census.gov/population/censusdata/urpop0090.txt.

5. Karl H. Oedekoven, "Forest History of the Near East," *Unasylva* 17, (Cairo: Forestry and Forest Products Division, Food and Agricultural Organization of the United Nations, 1963), http://www.fao.org/docrep/e3200e/e3200e03.htm.

6. Trees for the Future, "Trees for Livestock: Why and How Forest Gardens Must be Used to Improve Livestock Rearing Practices, Reverse Land Degradation, and Increase Smallholder Income," (internal white paper, Trees for the Future, Silver Spring, Maryland, 2015).

7. Trees for the Future, "Trees for Livestock."

8. Tim Robinson, "Livestock and livestock production in the African drylands," *Living Data* (blog), *CGIAR,* August 29, 2014, http://dialogues.cgiar.org/blog/livestock-and-livestock-production-in-the-african-drylands/.

9. Lester R. Brown, *World on the Edge: How to Prevent Environmental and Economic Collapse* (New York: W.W. Norton & Company, 2011).

10. Bill Forse, "The myth of the marching desert," *New Scientist,* no. 1650 (1989): 31-32.

11. Jim Morrison, "The 'Great Green Wall' Didn't Stop Desertification, But It Evolved into Something That Might," *Smithsonian Magazine,* August 23, 2016, http://www.smithsonianmag.com/science-nature/great-green-wall-stop-desertification-not-so-much-180960171/?utm_source=facebook.com&no-ist.

12. "FAOSTAT," Food and Agricultural Organization of the United Nations, queried by Benin, Burkio Faso, Chad, Eritrea, Ethiopia, Gambia, Mali, Mauritania, Niger, Nigeria, Senegal, Sudan, 1978-2008, cattle, sheep and goats, accessed September 26, 2016, http://www.fao.org/faostat/en/#data/QA

13. World Conservation Union, Norwegian Agency for International Development, Sahel Programme, *The IUCN Sahel Studies 1989* (Gland: IUCN, 1989). https://books.google.com/books?id=3jRmxGZhSt4C&printsec=frontcover#v=onepage&q&f=false.

14. Philip Thornton and Mario Harrero, "The Inter-linkages between Rapid Growth in Livestock Production, Climate Change, and the Impacts on Water Resources, Land Use, and Deforestation," Policy Research Working Papers, *The World Bank* (2010), doi: 10.1596/1813-9450-5178.

15. Thornton and Harrero, "Inter-linkages between Rapid Growth in Livestock Production, Climate Change, and the Impacts on Water Resources, Land Use, and Deforestation."

16. Nikos Alexandratos and Jelle Bruinsma, "World Agriculture Towards 2030/2050: The 2012 Revision," (ESA Working Paper No. 12-03, Agricultural Development Economics Division, Food and Agriculture Organization of the United Nations, Washington, DC, June 2012).

17. Thornton and Harrero, "Inter-linkages between Rapid Growth in Livestock Production, Climate Change, and the Impacts on Water Resources, Land Use, and Deforestation."

18. "The Taylor Grazing Act," Casper Office, Bureau of Land Management, U.S. Department of the Interior, last modified January 13, 2011, https://www.blm.gov/wy/st/en/field_offices/Casper/range/taylor.1.html.

19. GRAIN, *Land grabbing and food sovereignty in West and Central Africa*, Against the Grain (Barcelona: GRAIN, August 2012).

20. Richard Shawyer, *Wisdom of the Wolof Sages: A Collection of Proverbs from Senegal translated and explained in English* (n.p: privately printed, 2009).

21. Trees for the Future, "Technical Reference Guide: Draft," (internal document, Silver Spring, Maryland).

22. Trees for the Future, "Technical Reference Guide."

23. Trees for the Future, "Technical Reference Guide."

24. Trees for the Future, "Technical Reference Guide."

25. Robin Walter and Sebastian Tsocanos, "Overgrazing Is Dead," *Rediscovering the Great American Prairie* (blog), August 22, 2014, https://rediscovertheprairie.org/2014/08/22/overgrazing-is-dead/.

CHAPTER 4: HYDE'S SEED/JEKYLL'S SOIL

1. Peace Corps Volunteer, in discussion with John Leary (2001).

2. Patricia S. Muir, "History of Pesticide Use," Oregon State University: BI301 Human Impacts on Ecosystems, last modified October 22, 2012, http://people.oregonstate.edu/~muirp/pesthist.htm

3. Marie-Monique Robin, *The World According to Monsanto: Pollution, Corruption, and the Control of the World's Food Supply* (New York: The New Press, 2010).

4. Mark B. Tauger, *Agriculture in World History* (New York: Routledge, 2013).

5. *Food, Inc.*, paid YouTube video, 1:33:45, directed by Robert Kenner (New York: Magnolia Pictures, 2009), https://www.youtube.com/watch?v=jRp71BwRW8c&feature=youtu.be

6. Memorandum from Dr. Samuel I. Shibko, to Dr. James Maryanski 1 (January 31, 1992) (R. 18980), available from http://biointegrity.org/FDAdocs/03/03.pdf.

7. Memorandum from Director, Center of Veterinary Medicine, to Biotechnology Coordinator (February 5, 1992) (R. 18990-18995), available from http://www.biointegrity.org/FDAdocs/08/08.pdf.

8. Memorandum from Chairman, FDA Task Group on Food Biotechnology, to The Director, Center for Food Safety and Applied Nutrition 1 (August 15, 1991) (R. 18428), available from http://biointegrity.org/FDAdocs/22/JMtoCFSp.pdf

9. Memorandum from Mitchell J. Smith, Ph.D., to Jim Maryanski, 1 (January 8, 1992) (R. 18960), available http://biointegrity.org/FDAdocs/07/07.pdf.

10. Federal Food and Drugs Act of 1906, Pub. L. No. 59-384, § 7, 34 Stat. 768 (repealed in 1938 by 21 U.S.C. Sec 329 (a)).

11. Foods Derived From New Plant Varieties, 57 Fed. Reg. 104, (29 May 1992). *U.S. Food and Drug Administration.* Web.

12. *Food, Inc.*, directed by Robert Kenner.

13. Theresa Phillips, "Genetically modified organisms (GMOs): Transgenic crops and recombinant DNA technology," *Nature Education* 1, no. 1 (2008): 213.

14. Steven Drucker, *Altered Genes, Twisted Truth: How the Venture to Genetically Engineer Our Food Has Subverted Science, Corrupted Government, and Systematically Deceived the Public* (Salt Lake City: Clear River Press, 2015).

15. United States Department of Agriculture, *Genetically engineered varieties of corn, upland cotton, and soybeans, by State and for the United States, 2000-16* (Washington, DC: Economic Research Service, United States Department of Agriculture, 2016), https://www.ers.usda.gov/data-products/adoption-of-genetically-engineered-crops-in-the-us/.

16. Jorge Fernandez-Cornejo, Seth Wechsler, Mike Livingston, and Lorraine Mitchell, *Genetically Engineered Crops in the United States* (report summary) (Washington, DC: United States Department of Agriculture, September 2014).

17. Fernandez-Cornejo et al., *Genetically Engineered Crops in the United States.*

18. Mark Arax and Jeanne Brokaw, "No Way Around Roundup," *Mother Jones,* January/February 1997, http://www.mother-jones.com/politics/1997/01/no-way-around-roundup.

19. Fernandez-Cornejo et al., *Genetically Engineered Crops in the United States.*

20. Fernandez-Cornejo et al., *Genetically Engineered Crops in the United States.*

21. Saranyu Khammuang, Srisulak Dheeranupattana, Prasert Hanmuangjai, and Sasitorn Wongroung, "Agrobacterium-mediated transformation of modified antifreeze protein gene in strawberry," *Songklanakarin Journal of Science and Technology* 27, no. 4 (2005): 693-703.

22. Ned Potter and Paul Eng, "Spinning Tough Spider Silk from Goat Milk," *ABC News,* January 31, 2002, http://abcnews.go.com/Technology/CuttingEdge/story?id=98095&page=1.

23. Amanda Holpuch, "Scientists Breed Glow-in-the-Dark Rabbits," *The Guardian,* August 13, 2013, https://www.theguardian.com/world/2013/aug/13/glow-in-dark-rabbits-scientists.

24. Anne Minard, "Gene-Altered 'Enviropig' to Reduce Dead Zones?" *National Geographic,* March 30, 2010, http://news.nationalgeographic.com/news/2010/03/100330-bacon-pigs-enviropig-dead-zones/.

25. Rick Weiss, "USDA Backs Production of Rice with Human Genes," *Washington Post,* March 2, 2007, http://www.washingtonpost.com/wp-dyn/content/article/2007/03/01/AR2007030101495.html.

26. B. Hammond, R. Dudek, J. Lemen, and M. Nemeth, "Results of a 13-week safety assurance study with rats fed grain from glyphosate tolerant corn," *Food Chemical Toxicology* 42, no. 6 (2004): 1003-1014, doi: 10.1016/j.fct.2004.02.013.

27. Gilles-Eric Séralini, Emilie Clair, Robin Mesnage, Steeve Gress, Nicolas Defarge, Manuela Malatesta, Didier Hennequin, and Joël Spiroux de Vendômois, "Retracted: Long term toxicity of a Roundup herbicide and a Roundup-tolerant genetically

modified maize," *Food and Chemical Toxicology* 50, no. 11 (2012): 4221-4231, doi: 10.1016/j.fct.2012.08.005.

28. Gilles-Eric Séralini, Emilie Clair, Robin Mesnage, Steeve Gress, Nicolas Defarge, Manuela Malatesta, Didier Hennequin and Joël Spiroux de Vendômois, "Republished study: long-term toxicity of a Roundup herbicide and a Roundup-tolerant genetically modified maize," *Environmental Sciences Europe* 26, (2014): 14, doi: 10.1186/s12302-014-0014-5.

29. Séralini et al., "Republished study: long-term toxicity of a Roundup herbicide and a Roundup-tolerant genetically modified maize," 4221-4231.

30. Séralini et al., "Republished study: long-term toxicity of a Roundup herbicide and a Roundup-tolerant genetically modified maize," 4221-4231.

31. John Peterson Myers, Michael N. Antoniou, Bruce Blumberg, Lynn Carroll, Theo Colborn, Lorne G. Everett, Michael Hansen, Philip J. Landrigan, Bruce P. Lanphear, Robin Mesnage, Laura N. Vandenberg, Frederick S. vom Saal, Wade V. Welshons, and Charles M. Benbrook, "Concerns over use of glyphosate-based herbicides and risks associated with exposures: a consensus statement," *Environmental Health* 15, (2016): 19, doi: 10.1186/s12940-016-0117-0.

32. "The Safety of Genetically Modified Foods Produced through Biotechnology," *Toxicological Sciences* 71, no. 1 (2003): 2-8, doi: 10.1093/toxsci/71.1.2.

33. Testbiotech, *A Playground of the Biotech Industry?: Need for Reform at the European Food Safety Authority* (Munich: Testbiotech, 2012), http://www.testbiotech.org/en/node/736.

34. Testbiotech, *A Playground of the Biotech Industry?*

35. "Who Will Feed Us?" Text and poster: The Industrial Food Chain or the Peasant Food Webs?" ETC Group, submitted September 6, 2013, http://www.etcgroup.org/content/poster-who-will-feed-us-industrial-food-chain-or-peasant-food-webs.

36. "Who Will Feed Us?" ETC Group.

37. "We Are Bio," Biotechnology Innovation Organization, accessed February 11, 2017, https://www.bio.org/.

38. Rick Weiss, "Firms Seek Patents on 'Climate Ready' Altered Crops," *Washington Post,* May 13, 2008, http://www.washingtonpost.com/wp-dyn/content/article/2008/05/12/AR2008051202919.html.

39. "Should We Grow GM Crops?" FRONTLINE and NOVA, PBS, 2001, http://www.pbs.org/wgbh/harvest/exist/arguments.html.

40. "Pocket K No. 42: Stacked Traits in Biotech Crops," International Service for the Acquisition of the Agro-biotech Applications, accessed February 11, 2017, http://isaaa.org/resources/publications/pocketk/42/default.asp.

41. Fernandez-Cornejo et al., *Genetically Engineered Crops in the United States.*

42. Christopher Heath and Anselm Kamperman Sanders eds., *New Frontiers of Intellectual Property Law: IP and Cultural Heritage—Geographical Indications—Enforcement—Overprotection* (Portland: Hart Publishing, 2005).

43. Center for Food Safety, *Monsanto vs. U.S. Farmers* (Washington, DC: Center for Food Safety, 2005), http://www.centerforfoodsafety.org/reports/1401/monsanto-vs-us-farmers#.

44. ETC Group, *Who Owns Nature?: Corporate Power and the Final Frontier in the Commodification of Life* (Munich: ETC Group, November 2008), 100.

45. Purdue University Cooperative Extension Service, "Soils, Agriculture, and Environment," in *Indiana Soils* (West Lafayette: Purdue University, 2004), https://www.agry.purdue. edu/soils_judging/new_manual/ch3-crop.html.

46. Food and Agriculture Organization of the United Nations, *World fertilizer trends and outlooks to 2018* (Rome: Food and Agriculture Organization of the United Nations, 2015): 10, available from http://www.fao.org/3/a-i4324e.pdf.

47. International Plant Nutrition Institute, *Diammonium Phosphate,* Nutrient Source Specifics 17 (Norcross, Georgia: International Plant Nutrition Institute, n.d.), http://www. ipni.net/publication/nss.nsf/catalog?ReadForm&cat=D.

48. Craig Pittman, "Phosphate Giant Mosaic Agrees to Pay Nearly $2 Billion over Mishandling of Hazardous Waste," *Tampa Bay Times,* October 1, 2015, http://www.tampabay. com/news/environment/phosphate-giant-mosaic-agrees-to-pay-2-billion-over-mishandling-of/2247897.

49. Bill Gates, "Who Will Suffer Most from Climate Change," YouTube video, 1:44, September 1, 2015, https://www.youtube. com/watch?v=CyXyy8aFEqY&feature=youtu.be

50. Purdue University Cooperative Extension Service, "Soils, Agriculture, and Environment."

51. "Fighting Peak Phosphorus," Massachusetts Institute of Technology, accessed February 11, 2017, http://web.mit. edu/12.000/www/m2016/finalwebsite/solutions/phosphorus. html.

52. "IRS Form 990: Return of Organization Exempt From Income Tax," Alliance for the Green Revolution of Africa, *GuideStar,* 2007, https://www.guidestar.

org/FinDocuments/2007/980/513/2007-980513530-
050cf706-9.pdf.

53. Alliance for a Green Revolution in Africa, *Annual Report 2015*
(Nairobi: AGRA, 2015), https://agra.org/2015AnnualReport/
download/.

54. Justin Gillis, "Norman Borlaug, Plant Scientist Who Fought
Famine, Dies at 95," *The New York Times,* September 13,
2009, http://www.nytimes.com/2009/09/14/business/ener-
gy-environment/14borlaug.html.

55. "IRS Form 990: Return of Organization Exempt
From Income Tax," Alliance for the Green Revolution
of Africa, *GuideStar,* 2015, https://www.guidestar.
org/FinDocuments/2015/980/513/2015-980513530-
0ccdccc1-9.pdf.

56. Isabella Kenfield, "Michael Taylor: Monsanto's Man in the
Obama Administration," Organic Consumers Association,
August 14, 2009, https://www.organicconsumers.org/news/
michael-taylor-monsantos-man-obama-administration.

57. World Economic Forum, "What If the World's Soil Runs
Out?" *Time,* December 14, 2012, http://world.time.
com/2012/12/14/what-if-the-worlds-soil-runs-out/.

58. Peg Herring, "The Secret Life of Soil," Oregon State
University Extension Service, last modified February
2, 2010, http://extension.oregonstate.edu/gardening/
secret-life-soil-0.

59. Peg Herring, "Secret Life of Soil."

60. "What Is the Nutrient Value of Lost Organic Matter by
Erosion?" Mahdi Al-Kaisi, *Iowa State University Extension
and Outreach,* March 6, 2015, http://crops.extension.

iastate.edu/cropnews/2015/03/what-nutrient-value-lost-organic-matter-erosion.

61. Suzanne Simard, "How Trees Talk to Each Other," filmed June 2016, TED video, 18:19, https://www.ted.com/talks/suzanne_simard_how_trees_talk_to_each_other.

62. Simard, "How Trees Talk to Each Other."

63. Dan McKinney, "Trees Found to Communicate through Fungi [Avatar!]," YouTube video, 4:40, from the film *Do Trees Communicate*, produced by Julia Dordel, 2011, posted by "MushrooMetropolis," November 23, 2014, https://www.youtube.com/watch?v=AJWE3y2xdhQ&feature=youtu.be

64. M. Nube and R. L. Voortman, "Human micronutrient deficiencies: linkages with micronutrient deficiencies in soils, crops and animal nutrition" in *Combating Micronutrient Deficiencies: Food-based Approaches*, eds. Brian Thompson and Leslie Amoroso (Rome: Food and Agriculture Organization of the United Nations), 289.

65. World Economic Forum, "What If the World's Soil Runs Out?"

66. Dan Charles, "A Mixed Blessing," *National Geographic Magazine*, May 2013, http://ngm.nationalgeographic.com/print/2013/05/fertilized-world/charles-text.

67. Trees for the Future, "Technical Reference Guide: Draft," (internal document, Silver Spring, Maryland).

68. Purdue University Cooperative Extension Service, "Soils, Agriculture, and Environment."

69. Trees for the Future, "Technical Reference Guide."

70. MacDonald Dzirutwe, "Africa Takes Fresh Look at GMO Crops as Drought Blights Continent," *Reuters,* January

7, 2016, http://www.reuters.com/article/us-africa-gmo-idUSKBN0UL1UN20160107.

71. Food and Agriculture Organization of the United Nations, *World fertilizer trends and outlooks to 2018*.

CHAPTER 5: TOO LITTLE TOO MUCH TOO DIRTY

1. Scott Harrison, "The Last Walk for Water," *Medium*, February 19, 2014, https://medium.com/charity-water/the-last-walk-for-water-979160375b4a#.cq6shqoe7.

2. "The water in you," The USGS Water Science School, United States Geological Survey, last modified December 2, 2016, http://water.usgs.gov/edu/propertyyou.html.

3. Chaitanya Iyyer, *Land Management* (Global India Publications, 2009).

4. "FLOW for Love of Water," YouTube video, 1:22:49, a documentary directed by Irena Salina in 2008, posted by "ajvaughan3 Documentary Films," September 22, 2015, https://www.youtube.com/watch?v=n6nXCphwzcQ&feature=youtu.be

5. Steven Solomon, *Water: The Epic Struggle for Wealth, Power, and Civilization* (New York: Harper Perennial, 2011), 396.

6. Solomon, *Water: The Epic Struggle for Wealth, Power, and Civilization*, 388–393.

7. *Encyclopaedia Britannica*, s.v. "Lake Nassar," last modified June 11, 2010, https://www.britannica.com/place/Lake-Nassar.

8. *Encyclopaedia Britannica*, s.v. "Lake Nassar."

9. Solomon, *Water: The Epic Struggle for Wealth, Power, and Civilization*, 390.

10. Solomon, Water: The Epic Struggle for Wealth, Power, and Civilization, 390.

11. Solomon, Water: The Epic Struggle for Wealth, Power, and Civilization, 391.

12. Solomon, Water: The Epic Struggle for Wealth, Power, and Civilization, 396.

13. "The Grand Ethiopian Renaissance Dam Fact Sheet," International Rivers, January 24, 2014, https://www.internationalrivers.org/resources/the-grand-ethiopian-renaissance-dam-fact-sheet-8213.

14. E.G. Woldegebriel, "Ethiopia to step up role as regional clean power exporter," *Reuters,* May 13, 2015, http://www.reuters.com/article/us-ethiopia-energy-idUSKBN0NY1EK20150513.

15. Dessalegn Rahmato, *Land to Investors: Large-Scale Land Transfers in Ethiopia* (Utrecht: Land Governance for Equitable and Sustainable Development, 2011).

16. E.G. Woldegebriel, "Ethiopia to step up role as regional clean power exporter," *Reuters.*

17. "Grand Ethiopian Renaissance Dam Fact Sheet," International Rivers.

18. "FLOW For Love of Water," directed by Irena Salina.

19. "FLOW For Love of Water," directed by Irena Salina.

20. "FLOW For Love of Water," directed by Irena Salina.

21. "Hoover Dam: Frequently Asked Questions and Answers; The Colorado River," Bureau of Reclamation, U.S. Department of the Interior, last modified March 12, 2015, https://www.usbr.gov/lc/hooverdam/faqs/riverfaq.html.

22. Denise Fort and Barry Belson, *Pipe Dreams: Water Supply Pipeline Projects in the West* (New York: The Natural Resources Defense Council, June 2012).

23. Jeff Guo, "Agriculture Is 80 Percent of Water Use in California. Why Aren't Farmers Being Forced to Cut Back?" *Washington Post,* April 3, 2015, https://www.washingtonpost.com/blogs/govbeat/wp/2015/04/03/agriculture-is-80-percent-of-water-use-in-california-why-arent-farmers-being-forced-to-cut-back/?utm_term=.baad1e15ae36.

24. Drew Fitzgerald, "Data Centers and Hidden Water Use," *The Wall Street Journal,* June 24, 2015, http://www.wsj.com/articles/SB10007111583511843695404581067903126039290.

25. Jeremy Hoeffs, "Measurements for an Olympic Size Swimming Pool," last modified August 20, 2013, http://www.livestrong.com/article/350103-measurements-for-an-olympic-size-swimming-pool/.

26. "Product Gallery," Water Footprint Network, http://waterfootprint.org/en/resources/interactive-tools/product-gallery/.

27. Janie Har, "California's Population Grows to Nearly 39.1 Million," *The San Diego Union-Tribune,* December 16, 2015, http://www.sandiegouniontribune.com/sdut-californias-population-grows-to-nearly-391-million-2015dec16-story.html.

28. Joseph Serna and Veronica Rocha, "8 Dead, 350 Square Miles Burned, 300 Homes Destroyed Since June in Grim Beginning to California Fire Season," *Los Angeles Times,* August 5, 2016, http://www.latimes.com/local/california/la-me-ln-fire-season-20160804-snap-story.html.

29. Alemante Gebre-Selassie, "Ethiopia Is in a State of Emergency. The Tyrannical Government Must Go," *The Guardian*, October 12, 2016, https://www.theguardian.com/commentisfree/2016/oct/12/ethiopia-state-of-emergency-protests-corruption.

30. Sena Christian, "A Dry Future Weighs Heavy on California Agriculture," *High Country News,* February 22, 2016, http://www.hcn.org/issues/48.3/a-dry-future-weighs-heavy-on-california-agriculture.

31. Baba Umar, "India's Shocking Farmer Suicide Epidemic," *Al Jazeera,* May 18, 2015, http://www.aljazeera.com/indepth/features/2015/05/india-shocking-farmer-suicide-epidemic-150513121717412.html.

32. Terezia Farkas, "Why Farmer Suicide Rates Are the Highest of Any Occupation," The Blog, *Huffington Post,* July 23, 2014, updated September 22, 2014, http://www.huffingtonpost.com/terezia-farkas/why-farmer-suicide-rates-_1_b_5610279.html.

33. Nathan Halverson, "9 Sobering Facts about California's Groundwater Problem," *Reveal,* June 25, 2015, https://www.revealnews.org/article/9-sobering-facts-about-californias-groundwater-problem/.

34. Kurtis Alexander, "Central Valley Sinking Fast because of Groundwater Pumping," *SFGATE,* August 20, 2015, http://www.sfgate.com/bayarea/article/Central-Valley-sinking-fast-because-of-6453686.php.

35. Alexander, "Central Valley Sinking Fast because of Groundwater Pumping."

36. Charles Fishman, *The Big Thirst: The Secret Life and Turbulent Future of Water* (New York: Free Press, 2011).

37. "Chief Seattle's Letter to All," California State University Northridge, last modified October 9, 2007, http://www.csun.edu/~vcpsy00h/seattle.htm.

38. "FLOW For Love of Water," directed by Irena Salina.

39. "FLOW For Love of Water," directed by Irena Salina.

40. "FLOW For Love of Water," directed by Irena Salina.

41. "FLOW For Love of Water," directed by Irena Salina.

42. Erin Wayman, "The Top Ten Human Evolution Discoveries from Ethiopia," *Smithsonian.com,* October 10, 2012, http://www.smithsonianmag.com/science-nature/the-top-ten-human-evolution-discoveries-from-ethiopia-67871931/.

43. Rahmato, *Land to Investors: Large-Scale Land Transfers in Ethiopia.*

44. Peg Herring, "The secret life of soil," Oregon State University Extension Service, February 2, 2010, http://extension.oregonstate.edu/gardening/secret-life-soil-0.

45. Arjen Y. Hoekstra, Ashok K. Chapagain, Maite M. Aldaya, and Mesfin M. Mekonnen, *The Water Footprint Assessment Manual* (London: Earthscan, 2011), 189. http://production.wfp.fabriquehq.nl/media/downloads/TheWaterFootprintAssessmentManual_2.pdf.

46. Randy Creswell and Franklin W. Martin, *Dryland Farming: Crops & Techniques for Arid Regions,* ECHO Technical Note (North Fort Myers: ECHO, 1993).

CHAPTER 6: WRITTEN IN STONE

1. Mihai Andrei, "The Heaviest Living Organism in the World," *ZME Science,* February 9, 2015, http://www.zmescience.com/other/science-abc/heaviest-organism-pando-aspen/.

2. "Deforestation and Desertification: The Effect; Forest Holocaust," *National Geographic,* January 3, 2017, http://www.nationalgeographic.com/eye/deforestation/effect.html.

3. Krystal D'Costa, "A Story of Wood," *Anthropology in Practice* (blog), *Scientific American,* June 4, 2015, https://

blogs.scientificamerican.com/anthropology-in-practice/
a-story-of-wood/.

4. Derrick Jensen and George Draffan, *Strangely Like War: The
Global Assault on Forests* (Vermont: Chelsea Green Publishing,
2003), 9.

5. Rebecca Roberts and Prof. Jared Diamond, "History,
Environment, Politics, Make Haiti Poor," *Talk of the Nation*,
February 4, 2010, transcript and radio broadcast, 17:14, http://
www.npr.org/templates/story/story.php?storyId=123374267.

6. Nathan McClintock, "Agroforestry and Sustainable Resources
Conservation in Haiti: A Case Study" (working paper, North
Carolina State University, Raleigh, North Carolina, 2003),
https://projects.ncsu.edu/project/cnrint/Agro/resource_home.htm.

7. Rebecca Roberts and Prof. Jared Diamond, "History,
Environment, Politics, Make Haiti Poor."

8. McClintock, "Agroforestry and Sustainable Resources
Conservation in Haiti."

9. McClintock, "Agroforestry and Sustainable Resources
Conservation in Haiti."

10. Live Science Staff, "Scientists: Haiti's Wildlife Faces Mass
Extinction," *Live Science*, November 17, 2010, http://www.
livescience.com/8986-scientists-haiti-wildlife-faces-mass-ex-
tinction.html.

11. "Haitian Results," *Tracing African Roots: Exploring the Ethnic
Origins of the Afro-Diaspora* (blog), June 11, 2016, https://trac-
ingafricanroots.wordpress.com/ancestrydna/haitian-results/.

12. John C. Cannon, "'This Is Not an Empty Forest': Africa's Palm
Oil Surge Builds in Cameroon," Global Palm Oil, *Mongabay*,
March 30, 2016, https://news.mongabay.com/2016/03/

this-is-not-empty-forest-africas-palm-oil-surge-builds-in-cameroon/.

13. Jeremy Hance, "A Huge Oil Palm Plantation Puts African Rainforest at Risk," *Yale Environment 360,* September 12, 2011, http://e360.yale.edu/features/huge_oil_palm_plantation_puts_africa_rainforest_at_risk

14. Cannon, "'This is Not an Empty Forest.'"

15. Hance, "Huge Oil Palm Plantation Puts African Rainforest at Risk."

16. "Palm Oil," Indonesia Investments, last modified February 2, 2016, http://www.indonesia-investments.com/business/commodities/palm-oil/item166.

17. "Deforestation and Desertification: The Effect; Forest Holocaust," *National Geographic,* December 15, 2016, http://www.nationalgeographic.com/eye/deforestation/effect.html.

18. Fred Pearce, "Global Extinction Rates: Why Do Estimates Vary So Wildly?," *Yale Environment 360,* http://e360.yale.edu/features/global_extinction_rates_why_do_estimates_vary_so_wildly.

19. "The Extinction Crisis," The Center for Biological Diversity, December 6, 2016, http://www.biologicaldiversity.org/programs/biodiversity/elements_of_biodiversity/extinction_crisis/.

20. Alanna Mitchell, "The 1,300 Bird Species Facing Extinction Signal Threats to Human Health," *National Geographic,* August 26, 2014, http://news.nationalgeographic.com/news/2014/08/140825-bird-environment-chemical-contaminant-climate-change-science-winged-warning/.

21. Elizabeth Kolbert, *The Sixth Extinction: An Unnatural History,* 3rd ed. (New York: Henry Holt and Company, 2014), 107.

22. Kolbert, *The Sixth Extinction: An Unnatural History*, 108–110.

23. J.D. Walker and J.W. Geissman, *2009 Geologic Time Scale* (Colorado: The Geological Society of America, 2009). *http*://quaternary.stratigraphy.org/correlation/GSAchron09.jpg.

24. Joseph Stromberg, "What Is the Anthropocene and Are We in It?" *Smithsonian Magazine,* January 2013, http://www.smithsonianmag.com/science-nature/what-is-the-anthropocene-and-are-we-in-it-164801414/.

25. Stromberg, "What Is the Anthropocene and Are We in It?"

26. Stromberg, "What Is the Anthropocene and Are We in It?"

27. Mark Monroe, *Before the Flood,* directed by Fisher Stevens.

28. "Part I: Terrestrial Zoology," 1947–1951, Encyclopedia Arctica, vol. 3, http://collections.dartmouth.edu/arctica-beta/html/EA03-06.html.

29. "Big Five Mass Extinction Events," BBC Nature, last modified October 2014, http://www.bbc.co.uk/nature/extinction_events.

30. "What Is Agrobiodiversity?" (Factsheet, Food and Agricultural Organization of the United Nations, Sustainability Pathways, 2004).

31. "What Is Agrobiodiversity?"

32. "What Is Agrobiodiversity?"

33. "Should We Grow GM Crops?," FRONTLINE and NOVA, PBS, 2001, http://www.pbs.org/wgbh/harvest/exist/arguments.html.

34. Sylvia Earle, "My Wish: Protect Our Oceans," TED video, filmed February 2009, 18:16, https://www.ted.com/talks/sylvia_earle_s_ted_prize_wish_to_protect_our_oceans?language=en.

35. Damian Carrington, "The Anthropocene Epoch: Scientists Declare Dawn of Human-Influenced Age," *The Guardian*, August 29, 2016, https://www.theguardian.com/environment/2016/aug/29/declare-anthropocene-epoch-experts-urge-geological-congress-human-impact-earth.

36. Earle, "My Wish: Protect Our Oceans."

37. Thomas L. Friedman, *Hot, Flat, and Crowded* (New York: Farrar, Straus, and Giroux, 2008).

38. Friedman, *Hot, Flat, and Crowded*.

39. R. Malleson, S. Asaha, M. Egot, M. Kshatriya, E. Marshall, K. Obeng-Okrah and T. Sunderland, "Non-timber forest products income from forest landscapes of Cameroon, Ghana and Nigeria: an incidental or integral contribution to sustaining rural livelihoods?," *International Forestry Review* 16, no. 3 (2014): 261-277.

40. Laurie Clark, *Non-Timber Forest Products Economics and Conservation Potential*, Congo Basin Information Series 10 (Washington, DC: Central African Regional Program for the Environment, U.S. Forest Service, 2001)

41. Malleson et al., "Non-timber forest products income from forest landscapes of Cameroon, Ghana and Nigeria," 261-277.

42. Malleson et al., "Non-timber forest products income from forest landscapes of Cameroon, Ghana and Nigeria," 261-277.

43. Malleson et al., "Non-timber forest products income from forest landscapes of Cameroon, Ghana and Nigeria," 261-277.

44. "November 2014-CAMEROON: Louis Nkembi, ERUDEF Executive Director," International Union for Conservation of Nature, November 3, 2014, https://www.iucn.org/content/november-2014-cameroon-louis-nkembi-erudef-executive-director.

45. "Cameroon," Fauna & Flora International, October 15, 2016, http://www.fauna-flora.org/explore/cameroon/.

46. Malleson et al., "Non-timber forest products income from forest landscapes of Cameroon, Ghana and Nigeria," 261-277.

47. "Palm Oil Fact Sheet," Rainforest Action Network, November 20, 2016, http://www.ran.org/palm_oil_fact_sheet.

48. "Forests," World Resources Institute, January 4, 2017, http://www.wri.org/our-work/topics/forests.

49. Ivette Perfecto, John H. Vandermeer, and Angus Lindsay Wright, *Nature's Matrix: Linking Agriculture, Conservation and Food Sovereignty.* (London: Earthscan, 2009).

50. Perfecto, Vandermeer, and Wright, *Nature's Matrix.*

51. Perfecto, Vandermeer, and Wright, *Nature's Matrix.*

52. Kathy Chiapaikeo, "Author speaks about edible forest gardens," *University Wire,* March 1, 2016.

53. Perfecto, Vandermeer, and Wright, *Nature's Matrix.*

54. Perfecto, Vandermeer, and Wright, *Nature's Matrix.*

55. Malleson et al., "Non-timber forest products income from forest landscapes of Cameroon, Ghana and Nigeria," 261-277.

56. William Armand Mala, Julius Chupezi Tieguhong, Ousseynou Ndoye, Sophie Grouwels, and Jean Lagarde Betti, "Collective action and promotion of forest based associations on non-wood forest products in Cameroon," *Development in Practice* 22, no. 8, (2012): 1122-1134, doi: 10.1080/09614524.2012.714353.

57. Malleson et al., "Non-timber forest products income from forest landscapes of Cameroon, Ghana and Nigeria," 261-277.

58. Ettagale Blauer and Jason Laure, *Cultures of the World: Mali* (New York: Marshall Cavendish Benchmark, 2008).

59. "The World's Fastest Growing Cities and Urban Areas from 2006 to 2020," The City Mayors Foundation, accessed January 5, 2017, http://www.citymayors.com/statistics/urban_ growth1.html.

60. James Brooke, "In Africa, Stoking Stoves to Save the Trees," *The New York Times,* March 20, 1988, http://www.nytimes. com/1988/03/20/world/in-africa-stoking-stoves-to-save-the-trees.html.

CHAPTER 7: THE HUNGER BAR

1. Ruth Alexander, "Does a Child Die of Hunger Every 10 Seconds?" *BBC News,* June 18, 2013, http://www.bbc. com/news/magazine-22935692.

2. Krista Larson, "Hunger Stalking Children in Senegal," *The Washington Times,* May 30, 2012, http://www.washington-times.com/news/2012/may/30/hunger-stalking-children/.

3. "The World Summit on Food Security and the People's Forum: Different Approaches to Addressing Global Hunger" (Informational Memorandum, No. 81, The Academic Council on the United Nations System, Rome, Winter 2010), http://acuns.org/wp-content/uploads/2012/06/ WorldSummitFoodSecurityandPeople.pdf.

4. "Different Approaches to Addressing Global Hunger" (Informational Memorandum, No. 81, The Academic Council on the United Nations System, Rome, Winter 2010).

5. Roger Thurow, *The First 1,000 Days: A Crucial Time for Mothers and Children—And the World* (New York: Public Affairs, 2016).

6. "Agriculture's key figures," Momagri, Dec 15, 2016, http://www.momagri.org/UK/agriculture-s-key-figures/

With-close-to-40-%25-of-the-global-workforce-agriculture-is-the-world-s-largest-provider-of-jobs-_1066.html.

7. "The Nutrition Challenge in Sub-Saharan Africa" Working Paper No. 2012-012, Regional Bureau for Africa (New York: United Nations Development Programme, January 2012).

8. "Different Approaches to Addressing Global Hunger" (Informational Memorandum, No. 81, The Academic Council on the United Nations System, Rome, Winter 2010).

9. Jonathan Foley, "A Five-Step Plan to Feed the World," *National Geographic Magazine,* December 5,2016, http://www.nationalgeographic.com/foodfeatures/feeding-9-billion/.

10. H. Charles, J. Godfray and Tara Garnett, "Food security and sustainable intensification," *Philosophical Transactions of the Royal Society B: Biological Sciences* 369, no. 1639 (2014): 20120273, doi: 10.1098/rstb.2012.0273.

11. Iris Mansour, "How Access to Fresh Food Divides Americans," *Fortune,* August 15, 2013, http://fortune.com/2013/08/15/how-access-to-fresh-food-divides-americans/.

12. Mark Bittman, "Everyone Eats There," *The New York Magazine,* October 10, 2012, http://www.nytimes.com/2012/10/14/magazine/californias-central-valley-land-of-a-billion-vegetables.html.

13. Beth Hoffman, "The 'Food Desert' in the Heart of California's Farming Region," *Grist,* February 2, 2011, http://grist.org/article/food-2011-02-01-the-food-desert-in-the-heart-of-californias-farming-region/.

14. Nathan Vardi, "America's Richest Counties," *Forbes,* April 11, 2011, http://www.forbes.com/2011/04/11/americas-richest-counties-business-washington.html.

15. Roni A. Neff, Anne M. Palmer, Shawn E. Mckenzie, and Robert S. Lawrence, "Food Systems and Public Health Disparities," *Journal of Hunger & Environmental Nutrition* 4, no. 3-4 (2009): 282-314, doi:10.1080/193202409033 37041.

16. Miguel A. Altieri, "Agroecology, Small Farms, and Food Sovereignty," *Monthly Review* 61, no. 3, July-August, http://monthlyreview.org/2009/07/01/agroecology-small-farms-and-food-sovereignty/.

17. Kingsley Ighobor, "Wangari Maathai, the Woman of Trees, Dies," *Africa Renewal Online,* October 3, 2011, http://www.un.org/africarenewal/web-features/wangari-maathai-woman-trees-dies.

18. Tamara Hinson, "Sahel Drought in West Africa Leading to Crisis as Millions of Lives at Risk," *Metro,* August 3, 2012, http://metro.co.uk/2012/08/03/sahel-drought-in-west-africa-leading-to-crisis-as-millions-of-lives-at-risk-521534/.

19. "Food-A-Pedia," United States Departments of Agriculture, accessed January 15, 2017, https://www.supertracker.usda.gov/foodapedia.aspx.

20. "Food-A-Pedia," United States Departments of Agriculture.

21. Aaron Woolf, Curt Ellis, Ian Cheney, and Jeffrey K. Miller, "INDEPENDENT LENS | King Corn | Extended Clip | PBS," YouTube video, 20:15, clip from King Corn, directed by Aaron Woolf, 2006, posted by PBS, April 23, 2008, https://www.youtube.com/watch?v=jDurZc5Yr6c

22. Alliance for a Green Revolution in Africa, *Annual Report 2015* (Nairobi: AGRA, 2015), https://agra.org/2015AnnualReport/download/.

CHAPTER 8: WEATHER OR NOT

1. Democracy Now, *Democracy Now*, May 10, 2013, https://www.democracynow.org/2013/5/10/headlines.

2. "For the first time Earth's single-day CO2 tops 400 ppm," NASA, May 10, 2013, https://climate.nasa.gov/news/916/for-first-time-earths-single-day-co2-tops-400-ppm/.

3. "GISS Surface Temperature Analysis: Additional Analysis Plotter," Goddard Institute for Space Studies, National Aeronautics and Space Administration, last modified January 17, 2017, http://data.giss.nasa.gov/gistemp/graphs/customize.html.

4. Brian Kahn, "Earth's CO2 Passes the 400 PPM Threshold—Maybe Permanently," *Scientific American*, September 27, 2016, https://www.scientificamerican.com/article/earth-s-co2-passes-the-400-ppm-threshold-maybe-permanently/.

5. "Trends in Atmospheric Carbon Dioxide: Up-to-date weekly average CO^2 at Mauna Loa," Global Monitoring Division, Earth System Research Laboratory, National Oceanic and Atmospheric Administration, U.S. Department of Commerce, last modified February 4, 2017, https://www.esrl.noaa.gov/gmd/ccgg/trends/weekly.html.

6. Kahn, "Earth's CO2 Passes the 400 PPM Threshold—Maybe Permanently."

7. Aradhna K. Tripati, Christopher D. Roberts, and Robert A. Eagle, "Coupling of CO_2 and Ice Sheet Stability Over Major Climate Transitions of the Last 20 Million Years," *Science* 326, no. 5958 (December 4, 2009): 1394-1397, doi: 10.1126/science.1178296.

8. Mary Bagley, "Precambrian: Facts about the Beginning of Time," *Live Science*, May 3, 2016, http://www.livescience.com/43354-precambrian-time.html.

9. Graham Readfearn, "Carbon Dioxide's 400ppm Milestone Shows Humans Are Rewriting the Planet's History," *The Guardian*, May 20, 2016, https://www.theguardian.com/environment/planet-oz/2016/may/20/carbon-dioxides-400ppm-milestone-shows-humans-are-re-writing-the-planets-history.

10. "Trends in Atmospheric Carbon Dioxide: Full Mauna Loa CO_2 record," Global Monitoring Division, Earth System Research Laboratory, National Oceanic and Atmospheric Administration, U.S. Department of Commerce, accessed December 12, 2016, https://www.esrl.noaa.gov/gmd/ccgg/trends/full.html.

11. "Satellites witness lowest arctic ice coverage in history," European Space Agency, last modified September 14, 2007, http://www.esa.int/Our_Activities/Observing_the_Earth/Envisat/Satellites_witness_lowest_Arctic_ice_coverage_in_history.

12. "Decline of West Antarctic Glaciers Appears Irreversible," NASA Earth Observatory, last modified May 16, 2014, http://earthobservatory.nasa.gov/IOTD/view.php?id=83672.

13. John D. Sutter, "You're Making This Island Disappear," CNN, June 2015, http://www.cnn.com/interactive/2015/06/opinions/sutter-two-degrees-marshall-islands/.

14. Nell Greenfieldboyce, "Study: 634 Million People at Risk from Rising Seas," *Morning Edition,* March 28, 2007, transcript and radio broadcast, 3:53, March 28, 2007, http://www.npr.org/templates/story/story.php?storyId=9162438.

15. "Heavy Flooding and Global Warming: Is There a Connection?" Union of Concerned Scientists, December 10, 2016, http://www.ucsusa.org/global_warming/science_and_impacts/impacts/heavy-flooding-and-global-warming.html.

16. CNN, "India Heat Melted Street," CNN video, 0:54, May 25, 2016, http://www.cnn.com/videos/world/2016/05/25/india-heat-melted-street-orig-vstan.cnn.

17. Food and Agriculture Organization of the United Nations, *Climate change and food security: risks and responses* (Rome: Food and Agriculture Organization of the United Nations, 2016).

18. "Billion-Dollar Weather and Climate Disasters: Table of Events," NOAA National Centers for Environmental Information, 2017, https://www.ncdc.noaa.gov/billions/events.

19. Charles Graeber, "Canada's $6.9 Billion Wildfire Is the Size of Delaware—and Still Out of Control," *Bloomberg Businessweek*, June 2, 2016, https://www.bloomberg.com/features/2016-wildfire-fort-mcmurray/.

20. Tom Moore, "A Week-Long Siege of Heavy Rain Triggers Flash Flooding in Texas in May, June 2016," *The Weather Channel*, June 3, 3016, https://weather.com/storms/severe/news/flash-flooding-texas-severe-weather-forecast-plains-may27-0.

21. Allie Goolrick, "Disease Turning Florida Oranges Green, Killing Trees," *The Weather Channel*, June 16, 2014, https://weather.com/science/news/usda-funds-citrus-greening-research-florida-citrus-industry-crisis-20140615.

22. Stephanie Murray, "Drought Produces Western Mass. Region's Driest Summer Since 1956," *WWLP-22News*, August 29,

2016, http://wwlp.com/2016/08/29/drought-produces-western-mass-regions-driest-summer-since-1956/.

23. Jon Erdman, "America's Most Extreme Weather Cities of 2016; At Least 58 Locations Set or Tied a Record Warm Year," *The Weather Channel*, December 29, 2016, https://weather.com/news/climate/news/most-extreme-weather-us-cities-2016.

24. "An Inconvenient Truth," *IMDb*, July 15, 2016, http://www.imdb.com/title/tt0497116/.

25. Mark Monroe, *Before the Flood*, paid YouTube Video, directed by Fisher Stevens (Washington D.C.: National Geographic Society, 2016), https://www.youtube.com/watch?v=ohwU3Sfdckc&feature=youtu.be

26. "The Paris Agreement" (Key Steps, United Nations Framework Convention on Climate Change, Bonn, November 2016), http://unfccc.int/paris_agreement/items/9485.php.

27. John Cassidy, "A Skeptical Note on the Paris Climate Deal," *The New Yorker*, December 14, 2015, http://www.newyorker.com/news/john-cassidy/skeptical-note-paris-climate-deal.

28. Coral Davenport and Mark Landler, "U.S. to Give $3 Billion to Climate Fund to Help Poor Nations, and Spur Rich Ones," *The New York Times*, November 14, 2014, https://www.nytimes.com/2014/11/15/us/politics/obama-climate-change-fund-3-billion-announcement.html?_r=0.

29. United Nations Conference on Trade and Development, *Trade and Environment Review 2013: Wake Up Before It's Too Late* (Geneva: United Nations, 2013): 39, available from http://unctad.org/en/PublicationsLibrary/ditcted2012d3_en.pdf.

30. Jon Rynn, "A Detailed Look at Building, Industry, Transportation, and Land-use Greenhouse-Gas Emissions," *Grist*, January

16, 2009, http://grist.org/article/convenient-facts-about-an-inconvenient-truth-part-2/.

31. "Climate-Smart Agriculture" (Rome: Food and Agriculture Organization of the United Nations, 2017), http://www.fao.org/climate-smart-agriculture/en/.

32. Talia Schmitt, "The Debate Over 'Climate-Smart' Agriculture," Pulitzer Center on Crisis Reporting, April 26, 2016, http://pulitzercenter.org/reporting/debate-over-climate-smart-agriculture.

33. Ben Lilliston, "The Clever Ambiguity of Climate Smart Agriculture," *Think Forward Blog*, Institute for Agriculture and Trade Policy, December 4, 2015, http://www.iatp.org/blog/201512/the-clever-ambiguity-of-climate-smart-agriculture.

34. Eugene Takle and Don Hofstrand, "Global Warming—Impacts of Global Climate Change on the Midwest," Iowa State University Extension and Outreach, July 2008, https://www.extension.iastate.edu/agdm/articles/others/TakJuly08.html.

35. United Nations Conference on Trade and Development, *Trade and Environment Review*.

36. United States Department of Agriculture, *Table 5—Corn supply, disappearance and share of total corn used for ethanol* (Washington, DC: Economic Research Service, United States Department of Agriculture, October 2016), https://www.ers.usda.gov/data-products/us-bioenergy-statistics/us-bioenergy-statistics/#Feedstocks.

37. "Ethanol Fuel History," MLR Solutions—Fuel Testers Company, last modified 2009, http://www.fuel-testers.com/ethanol_fuel_history.html.

38. Jay Kimball, "Farming Wind Versus Farming Corn for Energy," 8020 Vision, September 2, 2010, http://8020vision.com/2010/09/02/farming-wind-versus-farming-corn-for-energy/.

39. Linda Hardesty, "It Takes 2.8 Acres of Land to Generate 1GWh of Solar Energy Per Year, Says NREL," Energy Manager Today, August 1, 2013, http://www.energymanagertoday.com/it-takes-2-8-acres-of-land-to-generate-1gwh-of-solar-energy-per-year-says-nrel-094185/.

40. John Cassidy, "A Skeptical Note on the Paris Climate Deal."

41. Peter Applebome, "They Used to Say Whale Oil Was Indispensable, Too," *The New York Times*, August 3, 2008, http://www.nytimes.com/2008/08/03/nyregion/03towns.html.

42. Applebome, "They Used to Say Whale Oil Was Indispensable, Too."

43. Rachael Petersen, Nigel Sizer, and Peter Lee, "Tar Sands Threaten World's Largest Boreal Forest," *Insights* (blog), World Resources Institute, July 15, 2014, http://www.wri.org/blog/2014/07/tar-sands-threaten-world%E2%80%99s-largest-boreal-forest.

44. Petersen, Sizer, and Lee, "Tar Sands Threaten World's Largest Boreal Forest."

45. Mark Monroe, *Before the Flood*, directed by Fisher Stevens.

46. "John F. Kennedy Moon Speech—Rice Stadium: September 12, 1962," NASA, last modified May 24, 2012, https://er.jsc.nasa.gov/seh/ricetalk.htm.

47. Arthur Neslen, "Wind Power Generates 140% of Denmark's Electricity Demand," *The Guardian*, July 10, 2015, https://www.theguardian.com/environment/2015/jul/10/denmark-wind-windfarm-power-exceed-electricity-demand.

48. Samuel Osborne, "Sweden Phases Out Fossil Fuels in Attempt to Run Completely Off Renewable Energy," *Independent*, May 24, 2016, http://www.independent.co.uk/news/world/europe/sweden-phases-out-fossil-fuels-in-attempt-to-run-completely-off-renewable-energy-a7047306.html.

49. "Number of mobile phone users worldwide from 2013 to 2019 (in billions)," Statista, accessed December 12, 2017, https://www.statista.com/statistics/274774/forecast-of-mobile-phone-users-worldwide/.

50. Jay Yarow, "Chart of the Day: More People Have Mobile Phones than Electricity or Drinking Water," *Business Insider*, April 30, 2012, http://www.businessinsider.com/chart-of-the-day-putting-global-mobile-in-context-2012-4.

51. "Diller Family History," ancestry, accessed December 3, 2016, http://www.ancestry.com/name-origin?surname=diller.

52. "Pacific Rim," *IMDb*, accessed October 15, 2016, http://www.imdb.com/title/tt1663662/.

53. "Ocean Acidification," The Ocean Portal Team, Smithsonian National Museum of Natural History, accessed November 10, 2016, http://ocean.si.edu/ocean-acidification.

54. Holli Riebeek, "The Ocean's Carbon Balance," *NASA Earth Observatory*, last modified June 30, 2008, http://earthobservatory.nasa.gov/Features/OceanCarbon/.

55. Judith D. Schwartz, "Soil as Carbon Storehouse: New Weapon in Climate Fight?" *Yale Environment 360*, March 4, 2014, http://e360.yale.edu/feature/soil_as_carbon_storehouse_new_weapon_in_climate_fight/2744/

56. Nell Greenfieldboyce, "Tree Counter Is Astonished by How Many Trees There Are," *Weekend Edition Saturday*,

December 26, 2015, transcript and radio broadcast, 3:54, http://www.npr.org/2015/12/26/461095807/tree-counter-is-astonished-by-how-many-trees-there-are.

57. Greenfieldboyce, "Tree Counter Is Astonished By How Many Trees There Are."

58. Dirk Bryant, Daniel Nielsen, and Laura Tangley, *The Last Frontier Forests: Ecosystems & Economies on the Edge* (Washington, DC: World Resources Institute, 1997).

59. "The incredible plan to make money grow on trees," *The Guardian*, November 24, 2015, https://www.theguardian.com/world/2015/nov/24/redd-papua-new-guinea-money-grow-on-trees.

60. "1998-Hurricane Mitch," Graduate School of Oceanography, The University of Rhode Island, accessed February 1, 2016, http://www.hurricanescience.org/history/storms/1990s/mitch/.

61. "Central American Leaders Urge Debt Relief," *BBC News*, November 10, 1998, http://news.bbc.co.uk/2/hi/americas/211228.stm.

CHAPTER 9: GREENER GRASS

1. Peace Corps Volunteer (independent consultant), in discussion with John Leary, 2005.

2. United Nations Department of Economic and Social Affairs, *World Urbanization Prospects: The 2014 Revision; Highlights* (New York: United Nations, 2014), available from https://esa.un.org/unpd/wup/publications/files/wup2014-highlights.Pdf.

3. United Nations Office on Drugs and Crime, *Global Report on Trafficking in Persons* (Vienna: United Nations, 2014), available

from http://www.unodc.org/documents/data-and-analysis/
glotip/GLOTIP_2014_full_report.pdf.

4. United Nations Office on Drugs and Crime, *Global Report on Trafficking in Persons.*

5. "Children HIV and AIDS," *AVERT,* last modified January 10, 2017, http://www.avert.org/professionals/hiv-social-issues/key-affected-populations/children.

6. United Nations Department of Economic and Social Affairs, *World Urbanization Prospects.*

7. Mediel Hove, Emmaculate Tsitsi Ngwerume, and Cyprian Muchemwa, "The Urban Crisis in Sub-Saharan Africa: A Threat to Human Security and Sustainable Development," *Stability: International Journal of Security and Development* 2, no. 1 (2013), doi: 10.5334/sta.ap.

8. Hove, Tsitsi Ngwerume, and Muchemwa, "The Urban Crisis in Sub-Saharan Africa."

9. Hove, Tsitsi Ngwerume, and Muchemwa, "The Urban Crisis in Sub-Saharan Africa."

10. "Urban Migration: Collapse Of Nigeria's Urban Infrastructure," 21:52, YouTube video, from an original news broadcast by Channels Television, posted by "Channels Television," May 17, 2013, https://www.youtube.com/watch?v=1CDV-GHdbAc&feature=youtu.be

11. "Urban Migration: Collapse Of Nigeria's Urban Infrastructure," original news broadcast by Channels Television.

12. Ward Anseeuw, Liz Alden Wily, Lorenzo Cotula, and Michael Taylor, *Land Rights and the Rush for Land: Findings of the Global Commercial Pressures on Land Research Project* (Rome: International Land Coalition, January 2012).

13. Anseeuw et al., *Land Rights and the Rush for Land.*

14. Karl Vick, "Libya's Migrant Economy Is a Modern Day Slave Market," *Time,* October 21, 2016, http://time.com/4538445/libyas-migrant-economy-is-a-modern-day-slave-market/.

15. Brad Plumer, "Drought Helped Cause Syria's War. Will Climate Change Bring More Like It?" *The Washington Post,* September 10, 2013, https://www.washingtonpost.com/news/wonk/wp/2013/09/10/drought-helped-caused-syrias-war-will-climate-change-bring-more-like-it/?utm_term=.b253bff5b020.

16. Jessica Barnes, "Managing the Waters of Bath Country: The Politics of Water Scarcity in Syria," *Geopolitics* 14, no. 3 (2009): 510-530, doi: 10.1080/14650040802694117.

17. Aron Lund, "Drought, Corruption, and War: Syria's Agricultural Crisis," *Diwan* (blog), April 18, 2014, http://carnegie-mec.org/diwan/55376.

18. Syrian Agriculture Database (Land & Water domain, Development of Wells by Governorate, Type, & License data collection; January 30, 2017), http://www.agriportal.gov.sy/napcsyr/sadb.htm.

19. Aron Lund, "Drought, Corruption, and War: Syria's Agricultural Crisis."

20. Brad Plumer, "Drought Helped Cause Syria's War. Will Climate Change Bring More Like It?"

21. Alexander de Waal and Human Rights Watch (Organization). *Evil Days: Thirty Years of War and Famine in Ethiopia.* Human Rights Watch, 1991.

22. de Waal, *Evil Days: Thirty Years of War and Famine in Ethiopia.*

23. Kevin Sieff, "A Smugglers' Haven in the Sahara," *Washington Post*, July 20, 2015, http://www.washingtonpost.com/sf/world/2015/07/20/a-remote-city-of-smugglers/.

24. United Nations High Commissioner on Refugees, *Global Trends: Forced Displacement in 2015* (Geneva: UNHCR, 2016), available from http://www.unhcr.org/576408cd7.pdf.

25. "U.S Immigration History," EIS research team, Progressives for Immigration Reform, accessed January 5, 2017, http://www.immigrationeis.org/about-ieis/us-immigration-history.

26. Hein de Haas, "Trans-Saharan Migration to North Africa and the EU: Historical Roots and Current Trends," *The Online Journal of the Migration Policy Institute*, November 1, 2006, http://www.migrationpolicy.org/article/trans-saharan-migration-north-africa-and-eu-historical-roots-and-current-trends.

27. Sieff, "A Smugglers' Haven in the Sahara."

28. Sieff, "A Smugglers' Haven in the Sahara."

29. United Nations Office on Drugs and Crime, *Global Report on Trafficking in Persons*.

30. Vick, "Libya's Migrant Economy Is a Modern Day Slave Market."

31. Kevin Sieff, "A Smugglers' Haven in the Sahara."

32. United Nations High Commissioner on Refugees, *Global Trends: Forced Displacement in 2015* (Geneva: UNHCR, 2016), available from http://www.unhcr.org/576408cd7.pdf.

33. United Nations High Commissioner on Refugees, *Global Trends: Forced Displacement in 2015*.

34. Nayla Rush, *Welcoming Unaccompanied Alien Children to the United States: Family Reunification Disguised as Refugee Resettlement* (Washington, DC: Center for

Immigration Studies, May 2016), http://cis.org/Welcoming-Unaccompanied-Alien-Children-to-the-United-States.

35. Rush, *Welcoming Unaccompanied Alien Children to the United States.*

36. Dario Lopez, Garance Burke, Frank Bajak, Alberto Arce, and Romina Ruiz-Goiriena, "U.S. Military Expands Its Drug War in Latin America," *USA Today,* February 3, 2013, http://www.usatoday.com/story/news/world/2013/02/03/us-expands-drug-war-latin-america/1887481/.

37. "Einstein forum aims to stem Africa brain drain," *Al Jazeera,* March 10, 2016, http://www.aljazeera.com/news/2016/03/einstein-forum-aims-stem-africa-brain-drain-160310095834426.html.

38. "Africa's Deadly Brain Drain—Malawi," YouTube video, 21:13, a video short produced by SBS/Dateline, November 2007, posted by Journeyman Pictures, January 2, 2008, https://www.youtube.com/watch?v=ME-ICeVKukA&feature=youtu.be.

39. Mfonobong Nsehe, "Meet the 35 Year-Old Entrepreneur Who Owns Nigeria's 2nd Largest Rice Farm," *Forbes,* June 27, 2016, http://www.forbes.com/sites/mfonobongnsehe/2016/06/27/meet-the-36-year-old-entrepreneur-who-owns-nigerias-2nd-largest-rice-farm/#25aa165771fd.

40. Nsehe, "Meet the 35 Year-Old Entrepreneur Who Owns Nigeria's 2nd Largest Rice Farm."

CHAPTER 10: CAN YOU SPARE SOME CHANGE?

1. "Over the Hedge," *IMDb,* accessed February 12, 2017, http://www.imdb.com/title/tt0327084/.

2. "Over the Hedge: Quotes," *IMDb,* accessed February 12, 2017, http://www.imdb.com/title/tt0327084/quotes.

3. *Food, Inc.,* paid YouTube video, 1:33:45, directed by Robert Kenner (New York: Magnolia Pictures, 2009), https://www.youtube.com/watch?v=jRp71BwRW8c&feature=youtu.be.

4. Nicholas Kristof, "Save the Darfur," *The New York Times,* May 10, 2007, http://www.nytimes.com/2007/05/10/opinion/10kristof.html.

5. "Organic Standards," Agricultural Marketing Service, United States Department of Agriculture, accessed February 12, 2017, https://www.ams.usda.gov/grades-standards/organic-standards.

6. "What Does Rainforest Alliance Certified™ Mean?" Rainforest Alliance, October 25, 2016, http://www.rainforest-alliance.org/faqs/what-does-rainforest-alliance-certified-mean.

7. Aaron Woolf, Curt Ellis, Ian Cheney, and Jeffrey K. Miller, "INDEPENDENT LENS | King Corn | Extended Clip | PBS," YouTube video, 20:15, clip from King Corn, directed by Aaron Woolf, 2006, posted by PBS, April 23, 2008, https://www.youtube.com/watch?v=jDurZc5Yr6c&feature=youtu.be.

8. "Six Degrees Could Change the World," YouTube video, 1:36:06, film directed by Ron Bowman, produced by National Geographic, 2008, posted by "Bogdan Iulian," June 24, 2013, https://www.youtube.com/watch?v=R_pb1G2wIoA&feature=youtu.be.

9. Cassandra Brooks, "Meat's Environmental Impact," Stanford Woods Institute for the Environment, July 25, 2011, https://woods.stanford.edu/news-events/news/meats-environmental-impact.

10. Nicholas Bakalar, "Risks: More Red Meat, More Mortality," *The New York Times,* March 12, 2012, http://www.nytimes.com/2012/03/13/health/research/red-meat-linked-to-cancer-and-heart-disease.html.

11. Mark Monroe, *Before the Flood,* paid YouTube Video, directed by Fisher Stevens (Washington D.C.: National Geographic Society, 2016), https://www.youtube.com/watch?v=ohwU3Sf-dckc&feature=youtu.be

12. Mark Monroe, "Before the Flood," directed by Fisher Stevens.

13. "Heartwarming Story about the Four Seasons of a Tree. Must Read!" Faye, *Elite Readers,* accessed February 12, 2017, https://www.elitereaders.com/heartwarming-story-about-four-seasons-of-a-tree/.

14. "Historical Timeline—Farmers & the Land," Growing a Nation, last modified 2014, https://www.agclassroom.org/gan/timeline/farmers_land.htm.

15. Carol Peppe Hewitt, *Financing Our Foodshed: Growing Local Food with Slow Money* (Gabriola Island: New Society Publishers, 2013).

16. Howard G. Buffett, *40 Chances: Finding Hope in a Hungry World* (New York: Simon and Schuster, 2013).

CHAPTER II: THE GUILDED AGE

1. Bill Gates, "Who Will Suffer Most from Climate Change?" YouTube video, 1:44, September 1, 2015, https://www.youtube.com/watch?v=CyXyy8aFEqY&feature=youtu.be.

2. Justin Gillis, "Norman Borlaug, Plant Scientist Who Fought Famine, Dies at 95," *The New York Times,* September 13, 2009, http://www.nytimes.com/2009/09/14/business/energy-environment/14borlaug.html.

3. Gillis, "Norman Borlaug, Plant Scientist Who Fought Famine, Dies at 95."

4. Gillis, "Norman Borlaug, Plant Scientist Who Fought Famine, Dies at 95."

5. Gillis, "Norman Borlaug, Plant Scientist Who Fought Famine, Dies at 95."

6. Vandana Shiva, *The Violence of the Green Revolution: Third World Agriculture, Ecology, and Politics*, (Lexington: The University Press of Kentucky, 2016), 34, https://muse.jhu.edu/.

7. "Groundhog Day," *IMDb,* accessed January 10, 2017, http://www.imdb.com/title/tt0107048/.

8. Nicolas Depetris Chauvin, Francis Mulangu, and Guido Porto, *Food Production and Consumption Trends in Sub-Saharan Africa: Prospects for the Transformation of the Agricultural Sector* (New York: United Nations Development Programme, 2012), available from http://www.africa.undp.org/content/rba/en/home/library/working-papers/food-production-con-sumption-trends.html.

9. United States Department of Agriculture, *Crop Production 2015 Summary,* (Washington, DC: National Agricultural Statistics Service, United States Department of Agriculture, January 2016), https://www.usda.gov/nass/PUBS/TODAYRPT/cropan16.pdf.

10. Megan Sheahana and Christopher B. Barrett, "Ten striking facts about agricultural input use in Sub-Saharan Africa," *Food Policy* (2016), doi: 10.1016/j.foodpol.2016.09.010.

11. "From Little Things Big Things Grow': The Story behind the Song," Australians Together, accessed February 12, 2017,

http://www.australianstogether.org.au/stories/detail/the-story-behind-the-song.

12. Heidi Dewan, "What Do YOU Stand For?" *Develop Daily* (blog), Develop Coaching, February 1, 2017, http://develop-coaching.com/blog/2017/2/1/what-do-you-stand-for-1.

CPSIA information can be obtained
at www.ICGtesting.com
Printed in the USA
BVHW031319140920
588785BV00001B/18